Charles Diehl, Emma Read Perkins

Excursions in Greece to recently explored sites of classical interest: Mycenae, Tiryns, Dodona, Delos, Athens, Olympia, Eleusis, Epidaurus, Tanagra. A popular account of the results of recent excavations

Charles Diehl, Emma Read Perkins

Excursions in Greece to recently explored sites of classical interest: Mycenae, Tiryns, Dodona, Delos, Athens, Olympia, Eleusis, Epidaurus, Tanagra. A popular account of the results of recent excavations

ISBN/EAN: 9783742875303

Manufactured in Europe, USA, Canada, Australia, Japa

Cover: Foto ©berggeist007 / pixelio.de

Manufactured and distributed by brebook publishing software (www.brebook.com)

Charles Diehl, Emma Read Perkins

Excursions in Greece to recently explored sites of classical interest: Mycenae, Tiryns, Dodona, Delos, Athens, Olympia, Eleusis, Epidaurus, Tanagra. A popular account of the results of recent excavations

EXCURSIONS IN GREECE.

Excursions in Greece

TO

Recently Explored Sites of Classical Interest:

*MYCENAE, TIRYNS, DODONA, DELOS, ATHENS, OLYMPIA,
ELEUSIS, EPIDAURUS, TANAGRA.*

A Popular Account of the Results of Recent Excavations.

BY

CHARLES DIEHL

(Late Member of the French Schools of Rome and Athens, and Professor at the
University of Nancy).

TRANSLATED BY

EMMA R. PERKINS

(Headmistress of the Girls' Grammar School, Thetford).

WITH AN INTRODUCTION BY

REGINALD STUART POOLE, LL.D.

(Professor of Archæology at University College, London; Corresponding Member
of the Institute of France).

WITH NINE PLANS AND FORTY-ONE ILLUSTRATIONS.

LONDON:
H. GREVEL AND CO.,
33, KING STREET, COVENT GARDEN, W.C.
NEW YORK: B. WESTERMANN & CO.
1893.

Printed by Hazell, Watson, & Viney, Ld., London and Aylesbury.

INTRODUCTION.

M. DIEHL has done a great service to archæological students. In the compass of a small volume he has given a clear summary of the recent discoveries in Greece. The list of contents is enough to show the largeness of the enterprise. He has only been able to accomplish it by a judicious acquaintance with the wants of the students. The matter is various, treating of no less than ten subjects: the range of time is vast, extending from the age of the earliest monuments to the fall of the Greek religion, a period of at least seventeen centuries, probably much more; yet the work is eminently satisfactory, marked by the French qualities of measure, form, and elimination. Those who desire to know more have only to refer to the list of works prefixed to each section, and after the judicious training afforded by the author they will have no difficulty in choosing the part most fitted to their special tastes. Few, we think, will attempt the whole circle of knowledge here so well planned.

Mycenae very properly begins the volume. It is

beyond question a very early site, perhaps the earliest, and it is the most interesting of its class. Its association with the romantic story of Schliemann enables the author to sketch that remarkable career, and in the later essays to deal with the discovery and not the discoverer. Otherwise the work would have been drawn out to an undue length. Yet in every case due honour is paid to the explorer of each site, whatever his nationality.

The probable date of the remains at Mycenae and the main results of the excavations are stated with a wise caution. Nowhere is the reader better prepared to form an independent judgment of a very difficult question. The approximative age, and the plan of the buildings on the Acropolis are not very difficult to grasp with the aid of the similar works at Tiryns. The problem which still awaits solution is the origin of the art, at once simple and advanced, which marks the structures of Mycenae, Tiryns, and Orchomenus. Two sources seem clear, one Asiatic, coming by land, and the other Egypto-Phoenician, coming by sea. As the products of these sources are intermixed, it is obvious that the date of either must direct us towards that of the other. It has not escaped M. Diehl that the tribute of the Kefta or Phoenicians to the Egyptian king Thothmes III. comprehends vases strikingly like those of Mycenae and other objects more analogous, as the bulls' heads of precious metal. In the accompanying hieroglyphic inscription these presents are said to be brought by the Phoenicians of the islands of the Mediterranean. The style is not purely Egyptian, it is more varied and more fanciful; but this in no way perplexes the question. The Phoenicians were never original. They adopted and gave their colour to the successive local styles of the Eastern Mediterranean; and from B.C. 1500, the age of Thothmes III., for about five centuries, the Egypto-Phoenician was the style of portable objects and wholly of decoration. At the same time nothing forbad the transport of purely Egyptian works easy to carry, and even of small objects brought to Egypt from the farther East. What is needed for a clearer solution of this question is a closer study of the Egyptian representation of the products of Phoenician art from B.C. 1500 downwards.

At Tiryns we see a fuller illustration of the architecture and art of the age of Mycenae than the greater citadel affords. The works of defence of the Acropolis and the decorations of the palace are in far better preservation; and here we clearly have the influence of the Phoenicians, both as builders and as decorators, and this, to a large extent, puts aside the influence of Asia Minor. The evidence of Orchomenus points in the same direction. Decorators skilled in Egyptian style must have worked at both sites. As a commentary on the Odyssey, Mycenae, Tiryns, and Orchomenus are of the highest value. An almost new branch of inquiry should here be started. We have at Tell-el-Amarneh, in Upper Egypt, an example of an Egyptian high-priest's palace of the same age, represented in a tomb of the place, not indeed fortified, yet constructed with the same object as a great dwelling-place, and remarkably similar in the main plan. Thus the Phoenician influence in Greece has again an Egyptian origin. Had the old Egyptian palaces within forts been discovered and planned, it is probable that the analogy would be complete.

The transition from the citadels with their palaces of the heroic ages to the oracle of Dodona seems violent. This is due to the great catastrophe of the old civilisation, which has made it hard for scholars to admit a gap in the documents which must be faced, and to see in the new and more primitive art the real parent of the wonders of the fifth century. But all that lies behind the actual remains at the Epirote Sanctuary is really very archaic. Thus Dodona is a survival of a very early worship, and a necessary commentary on the primitive palaces. The oracle, again, and the methods of consulting it carry us back to the heroic ages.

In the chapter on the excavations at the Acropolis of Athens, the author examines the late important discoveries of the statues of the old Parthenon broken during the Persian occupation of the city and used as mere material for the support of the new building. The character of the art before Pheidias is now clear; and the whole early movement of Greek sculpture may be traced, in remarkable justification of the date to which the vase-painting formerly assigned to the age of the Parthenon, is now generally carried back. This may be called the pre-Raphaelite period of Greek art, which will be more and more appreciated as its value in illustration of the succeeding bloom is more clearly perceived. We must no longer imagine the art of the Parthenon to have sprung into existence in full beauty as in the birth of Athena. That which the coins clearly tell us is now equally seen in the art to which they owed their origin.

The excavations in Delos, if less startling than others in their results, form a singularly complete chapter in the story of the restoration of the Hellenic past. The island is an example of a tract without vegetation, and with the solitary advantage of a port, which owed its fortune to the temple of Apollo and its surroundings, and as a place of pilgrimage protected by its sanctity won great commercial power—the Mekkeh of the Greek islands. In an admirable chapter the reader can study the long-continued and complete organisation, which gave Delos a many-sided importance, only to fall by the violence of Mithradates and the growth of the Roman ports of Italy.

The temple of Apollo Ptoïos in Bœotia is the

subject of a brief chapter, awaiting the publication of the large work of M. Holleaux the explorer.

A large space is given to Olympia, the most fruitful, because practically the most central site in Greece. No one can read Pindar, or study any branch of Greek antiquities, without seeing the widespread importance of the Olympic games and the place they held in Greek life wherever it was planted. In Macedon on the north, in Sicily on the west, in Cyrene on the south, they are alike commemorated more than in central Greece, Elis alone excepted. The treasuries also were mostly of the more remote cities, Syracuse, Metapontum, Sybaris, Selinus, Gela, and Megara. This is at first sight perplexing, for one would have expected rather that the cities of Greece Proper should be represented. Looking, however, at Pindar, and the coins, we note that these cities won and commemorated victories in the chariot race, which needed the wealth of their *tyrannoi* and in their commemoration reflected honour on the seats of the rulers by whose power they were won. It must be remembered that the *tyrannoi* did not personally contend, but were at the cost of sending their chariots and charioteers to the great contest. Though the presence of women was forbidden, they also could in this manner win a victory by deputy.

An excellent account is here given of the Olympic Games, on which the excavations have thrown great light. The knowledge of the site and of the position of the courses removes the contest from the circle of mere book-knowledge into that of the practical. No doubt there was a certain savage element in the personal combats not wanting wholly in the chariot-races, but it must be remembered that the physical training of Olympia hardened the Greek for war, and its decline was the presage of his loss of pre-eminence. The element of fierceness is not wanting in the practical picture here represented, but at its height it is far superior to the barbarism of the Roman gladiatorial shows. And the reference by St. Paul shows that the contest had a truly noble side in its courage, its endurance, its desire of glory alone.

The sculptures of Olympia form an archæological puzzle, here stated but not resolved. Those of the pediments of the Temple of Zeus have been to a large extent discovered, and it is quite possible to judge of their style. Individual statues have also rewarded the explorer, two of which are of the highest merit, the Nike of Paionios, and the Hermes of Praxiteles. The pedimental sculptures which were contemporary with those of the Parthenon and anciently attributed to Alcamenes and Paionios, are undoubtedly disappointing. To account for their inferiority to the marbles of the Parthenon two theories may be advanced. Either they were of earlier execution, dating from soon after the foundation of the temple in about B.C. 470, or the execution of the designs was mostly entrusted to local artists. There is a want of that unity of design which marks the sculptures of the Parthenon, and makes it hard to discriminate between the work of Pheidias in the pediments, and that of his students in the frieze, although the different hands in the metopes assert themselves. It is here alone in the Parthenon that a link may be traced with the sculptures of Olympia. It seems surprising that when Pheidias, called here to execute the crowning work of the temple, the gold and ivory statue of Zeus, built a workshop in which he could produce it with every advantage of light and position, yet the people of Olympia were content with the inferior art of the pediments; but we look in vain for an instance of sculptures removed for the purpose of replacing them by finer works, and it must be recollected that such a labour would have retarded other and more pressing enterprises, above all, the completion of the great statue.

To turn from Olympia to Eleusis is to go from sunlight into the dusk. The sanctuary indeed has been discovered, and traces of the neighbouring temples, but its purpose remains obscure. The Eleusinian mysteries remain true to their name. The ancients preserve a religious silence, and the early Fathers of the Church cannot be expected to have known much of that which was guarded with a masonic secrecy. At the same time the main outlines of the purpose of the mysteries, and their mode of celebration, are known to us, and it seems that they had no barbarous or foreign element, but are reflected though not revealed in the literature which they originated.

The excavations at Epidaurus have restored to us a special and very interesting phase of Greek belief. The idea of a healing power, sometimes apart from that of the physician, and sometimes associated with it, but never in opposition, lies at the root of the worship at Epidaurus. The Asclepieion was rather a hospital than a temple. Although it had the services of a temple with a regular priesthood, yet it was frequented by worshippers who came to be healed, and who remained for various periods lodged in the porticoes of the temple awaiting dreams and undergoing various kinds of treatment usually of an elementary character, far inferior to that of contemporary medicine and surgery. It is characteristic of the Greek instinct that birth and death were rigorously excluded from the temples and the sacred precincts which surrounded them. Therefore a poor sufferer might come from the ends of the Greek world only to perish of exhaustion at the threshold of the temple, or to be pitiably carried away at the last moment. It is a great honour to the memory of Antoninus Pius that he built outside the peribolos a great house where the dying could pass away in peace. This is a curious contrast between the humanity of the Greeks and that of the Romans, and a startling illustration of the opposite ideas of the Hellenes and the Egyptians.

Tanagra, and indeed Boeotia generally, has in our times strangely reversed the judgment of antiquity, by which the stupidity of the Boeotians passed into a proverb, in spite of Pindar, Corinna, Epaminondas, and the lesser fame of Plutarch. Throughout the country a multitude of terra-cottas have been discovered similar to those from all parts of the Greek world, yet clothed with a special charm which renders them far superior to those even of Italy and of Asia Minor.

The question of the origin of these terra-cottas has never been proposed. How did this new art arise? In the bloom of Greek art, youth and maturity are the periods chosen by the sculptor, and the preference is given to man. In these terra-cottas, of which the finest are undoubtedly of the beginning of the decline, after Alexander, girlhood and womanhood are the favourite themes, and all ages are represented. Here are the marks of a social change. If we call in the aid of history we see that the success of Alexander brought the Greek woman out of the oriental seclusion which marked Athenian life into a publicity consistent with modern ideas in the West. The kings had courts, and the nobles imitated their manners. A court made a queen necessary; almost immediately that queen became the sole wife, and if she had political skill her life was not alone public, but political, and even military. Hence there arose a series of queens unrivalled in history, because they were on their trial. Phila, the admirable wife of Demetrius Poliorcetes, and the long series of Egyptian queens, the two great Berenices, the second and third Arsinoës, the first and last Cleopatras, will occur to every one as women of capacity and even of genius, and in most cases of exemplary character. This social revolution seems to be the key to a fresh bloom of art. As the school of Lysippus inaugurated the new heroic style of sculpture, so the terra-cottas mark a new epoch in the representation of women and their cycle. Nothing could be more striking than the comparison of the infant Dionysus carried by Hermes, so unworthy of the great statue of Praxiteles that some critics have attributed it to another hand, and the delightful children of the terra-cottas beginning little more than a century later.

The actual purpose of the terra-cottas is more difficult to determine than their origin. Originally they were no doubt sepulchral, but it is hard to associate the finest with a purely sepulchral purpose. It seems better to consider them as primarily portraying everyday life, whatever their use in tombs. At least there is a curious analogy with the Italian art of the Renaissance. It begins with the most sincere religious feelings, and passes into a pure artistic love of beauty, wholly apart in many cases from the earlier instinct, which ultimately disappears. Whatever the purpose of the terra-cottas their connection with sepulture is beyond doubt, and the stronger sense of the reality of the future state may account for the representation of the life on this side of the tomb as the truest type of the life beyond. Here certainly the influence of Egypt would have been powerful if it touched the Greek mind, which in the decay of Hellenic belief was strongly attracted by any other form of faith, especially if it was concerned with the great mysteries of the future. It may be however that Eleusis is really the parent of the terra-cottas.

REGINALD STUART POOLE.

PREFACE.

IT was M. Diehl's wish, as he expressly states in his preface, to popularise a knowledge of the results of recent excavations in Greece, and especially to draw attention to their importance as throwing light on Greek manners, religion, and history.

It has been my hope, in undertaking the translation, that his book in its English form might introduce some readers to the study of classical archæology, and give them a portion of the pleasure which that study once gave me. A few notes have been added and a good many references.

I have to thank my sister, Miss S. R. Perkins, for her help in the translation; Professor Stuart Poole for his great kindness in writing the Introduction; and Mr. E. A. Gardner, Director of the British School of Archæology at Athens, for advice and criticism on several points; and to express my sense of the value of the permission to read in the Library of the Fitzwilliam Museum of Archæology, Cambridge, granted me some time ago by the kindness of the Director, Professor Middleton.

EMMA READ PERKINS.

THE SCHOOL HOUSE, THETFORD,
Sept. 28th, 1892.

CONTENTS.

INTRODUCTION . . . . page v

PREFACE . . . . „ xvii

CHAPTER I.

THE EXCAVATIONS AT MYCENAE (1876-88).

I. Schliemann's Autobiography—His Early Life—Acquisition of Wealth—Scientific Undertakings—Excavations at Hissarlik and Discovery of Troy . . . . . . pp. 2—7

II. Position and Ruins of Mycenae—Its Historical Associations—Excavations on the Acropolis—Discovery of Six Graves—Ornaments contained in them—Tomb of Agamemnon—Importance of Discoveries made at Mycenae . pp. 7—22

III. Different Theories regarding Civilisation of Mycenae—Excavations of 1887 and 1888—Different Elements of this Civilisation—Oriental Influence in Early Greece—Objects imported from the East—Objects of Home Manufacture at Home—Mycenaean Style—From what People was this Civilisation derived? . . . . . pp. 22—37

IV. Analogies between Mycenaean Objects and those discovered at Hissarlik, Santorin, Tiryns, Spata—Date and Characteristics of Mycenaean Civilisation . . pp. 37–40

CONTENTS.

CHAPTER II.

EXCAVATIONS AT TIRYNS (1884-85).

Position and Ruins of Tiryns — Schliemann's Excavations — Their Importance.

I. The Citadel at Tiryns—Walls—Subterranean Galleries—Gates — Royal Palace—Men's Apartment—Women's Apartment—Use of Wood in the Building—Frescoes—Comparison between this and the Homeric Palaces . . pp. 46—59

II. Comparisons between Palaces of Tiryns and Mycenae—Oriental Influence at Tiryns—Date of Tirynthian Civilisation—Its Duration and Downfall . . . pp. 59—62

CHAPTER III.

EXCAVATIONS AT DODONA (1876).

I. Epirus and the Valley of Dodona—Position and Ruins of the Sanctuary — Excavations of M. Carapanos — Chief Results . . . . pp. 66—68

II. Origin of the Cultus of Dodona—Priestly Corporation of Dodona — Asceticism in Antiquity — Oracle of Zeus — Questions addressed to it—Methods of Consultation—Leaden Plates with Inscriptions—Offerings of the Worshippers . . . . pp. 69—78

III. Clients of the Oracle—Fame and Reputation of the Shrine—Its Decadence—Imagines of Philostratus . . pp. 78—82

CHAPTER IV.

EXCAVATIONS ON THE ACROPOLIS OF ATHENS (1882-89).

The Early School of Attic Sculpture — Importance of Recent Excavations.

I. Discoveries made in 1863—Finds of 1882—Excavations of 1885—Discovery of Fourteen Female Statues—Their Date — Costume and Coiffure — Subjects Represented — Why they were buried in Ancient Times—Works undertaken by Cimon on the Acropolis—Pediment Sculptures in Tufa—Group of Heracles and Typhon—Polychromy in the Pediments — Later Discoveries of Female Statues — The Acropolis in the Time of Pisistratus—The Old Temple of Athens . . . . pp. 87—106

II. The Beginnings of Greek Sculpture—Creation of an Artistic Type—Its Successive Transformations—The Series of Female Figures—The Male Figures—The Type of Athena . . . . pp. 106—119

III. Various Sculptors who worked on the Acropolis—The Works of the Samian School—The Early Works of Native Sculptors—The School of Chios and its Influence—The Development of Attic Art . . . . . pp. 119—127

CHAPTER V.

THE EXCAVATIONS AT DELOS (1873-88).

I. Position of Delos—Causes of its Ancient Splendour—The Worship of Apollo—Trade and Commerce—The Festivals of Delos—History of the Excavations—The Explorations of M. Homolle—Appearance of the Ruins previous to Excavations—Inadequacy of Literary Sources—Importance of the Discoveries made . . . . . pp. 130—142

II. Topography of Delos—General Appearance—The Sacred Port—Gates—Temenos—Temple of Apollo—Sanctuary of the Bulls—Temenos of Artemis—Merchants' Quarter—Schools of the Italians—Portico of Philip—Mount Cynthus . . . . pp. 142—150

III. History of Delos—The Attico-Delian Confederation and the Athenian Rule—Delos in the Fourth Century B.C.—The Confederation of the Cyclades—Influence of Rhodes—Protectorates of Egypt, Syria, and Macedonia—Prosperity and Aims of Delos in Third Century B.C.—The Romans in Delos—Delos a Free Port—Commercial Relations of the Island—The Trading Guilds—The Port—The Downfall of Delos . . . . pp. 150—170

IV. The Administration of the Sacred Treasure—The Hieropoioi—Accounts and Inventories of the Offerings—The Revenues of Apollo—Leases and Loans—Expenditure—Public Works and Expenses of Public Worship—The Offerings—The Inventory of the Treasure—The Plate and Wardrobe of Apollo—Works of Art . . . pp. 170—185

V. The Marble of Delos—The Archaic School of Naxos—The School of Chios—The Statue of Archermus—Archaic Statues of Artemis—Pediments of the Temple of Apollo . pp. 185—193

CHAPTER VI.
EXCAVATIONS AT THE TEMPLE OF APOLLO PTOÏOS (1884-88).

Ruins and History of the Temple of Apollo Ptoïos—The Ptoïan Games—Importance of M. Holleaux's Excavations—The Archaic Type of Apollo—Transformed by the Bœotian School — Successive Modifications of the Type — The Statues of Perdicovrysi—Qualities of the Bœotian School —Another Archaic Type of Apollo—The Bronzes of the Temple of Apollo Ptoïos—The Apollo of Canachus and Replicas . . . . pp. 194—209

CHAPTER VII.
EXCAVATIONS AT OLYMPIA (1875-81).

I. Position of Olympia—Origin of the Olympian Games—Their Importance in Antiquity—Development and Splendour of the Festival—Ruins of Olympia—History of the Excavations—Montfaucon and Winckelmann—The Expedition of the Morea—The German Excavations—The Museum of Olympia . . . . pp. 210—225

II. Topography of the Sanctuary—The Country—The Roads—The Gates of the Altis—The Temple of Zeus—The Heræum—The Treasuries—Buildings in the Altis—Stadium and Hippodome—Leonidaeum . . pp. 225—235

III. The Olympian Games—Their Gradual Development—Preparation for the Festival—The Hellanodicae—The Competitors—The Spectators—The Foot-race—The Wrestling—The Boxing—The Pancratium—The Chariot Race—The Pentathlum—The Victors—The Offerings and the Statues . . . . pp. 235—254

IV. The Temple of the Olympian Zeus—Its Pediments—Description of the Scene represented—The Style of the Marbles of Olympia—Inequality between Composition and Execution—Paeonius and Alcamenes—The Victory of Paeonius—The Metopes of the Temple—The Forerunners of Phidias—The Statue of the Olympian Zeus . . . pp. 255—280

V. The Heræum—Its Construction—Use of Wood and Terracotta in Early Greek Architecture—Archaic Bronzes from Olympia—Oriental Influence—Pediment of the Treasury of Megara—The Hermes of Praxiteles—Qualities and Style of Praxiteles . . . . pp. 280 293

CHAPTER VIII.

EXCAVATIONS OF ELEUSIS (1882-89).

The Hymn to Demeter and the Myth of Eleusis—The Institution of the Mysteries.

I. Pious Memories attached to Eleusis—History and Buildings of the Sanctuary—The Buildings of Pericles—Eleusis in the Fourth Century and under Roman Rule—The Excavations at Eleusis . . . . pp. 300—308

II. The Sacerdotal Body at Eleusis—The Administration of the Temple—Part taken by Athens in the Eleusinian Cultus—The Mysteries—Our Information concerning them—The Lesser Mysteries—The Mystae and the ἐπόπτεια—The Greater Mysteries—The Miracle-play of Eleusis—Impression produced by these Sights—Character and Influence of the Eleusinian Mysteries . . . . . pp. 308—323

III. Topography of the Sanctuary—The Hall of Initiation—The Archaic Statues discovered at Eleusis—The Eubuleus of Praxiteles . . . . pp. 323—330

CHAPTER IX.

THE EXCAVATIONS AT EPIDAURUS (1881-87).

I. Character of the Cultus of Asclepius—Plan of the Sanctuaries of the God—The Buildings of Epidaurus—The Temple—The Tholus of Polycleitus—The Theatre—The Sacerdotal Body—Nature of Religious Medicine—Lay Medicine in Antiquity . . . . pp. 332—340

II. Religious Therapeutics at Epidaurus—Tablets recording Miraculous Cures—The Fees of the Physician-god—Asclepius and the Sceptics—Wide Diffusion of the Worship of Asclepius—One Day of a Patient—The Prescriptions of the God—The Offerings of the Faithful . . pp. 340—354

III. The Marbles of Epidaurus—The Pediments of the Temple—Decadence of the Sanctuary—Epidaurus and Scientific Medicine . . . . pp. 354—357

CHAPTER X.

EXCAVATIONS AT TANAGRA (1870-89).

I. Evil Reputation of Boeotia—Situation and History of Tanagra—Excavations at Tanagra—High Reputation of the Statuettes—Shape of the Tombs—Their Contents—Diversity of Terra-cottas found—The Figurines stamped on Flat Cakes—The Stamped Busts—The Statuettes of the Fourth Century . . . . . . pp. 359—369

II. Methods of Manufacture—The Moulds—How the Coroplast retouched and transformed the Figurine—Its Colouring—Industrial Character of this Manufacture . . pp. 369—375

III. What do the Tanagra Figurines represent?—M. Heuzey's Opinion—Are the Figurines Images of Divinities?—Genre Subjects—Popular and Industrial Nature of the Coroplast's Art—Why were Figurines placed in the Tomb?—Greek Conceptions of the Life beyond the Grave—Did the Statuettes serve to protect the Dead?—Were they to afford him Company in the Tomb?—Use made of the Statuettes by the Ancients—The Figurines have not an especially Sepulchral Character—They are Offerings to the Dead . . . . pp. 375—391

IV. Everyday Life depicted in the Figurines—Male Figures—The Child—The Ephebus—Female Life—Diversity of Figurines—The Young Girl—Education—Games—The Woman—Toilette—Costume—Coiffure . . pp. 391—401

EXCURSIONS IN GREECE.

CHAPTER I.

MYCENAE.

Books of Reference:
  Schliemann, *Mycenae.*
  Schuchhardt, *Schliemann's Excavations.*
  Furtwängler and Loeschke, *Mykenische Thongefässe.*
  Furtwängler and Loeschke, *Mykenische Vasen.*
  Dumont & Chaplain, *Les Céramiques de la Grèce propre.*
  Rayet and Collignon, *Histoire de la Céramique Grecque.*
  For Spata, see *Bull. de Corr. Hell.*, vol. ii.
  For Menidi, see *Das Kuppelgrab von Menidi.* Athens, 1880.
  For Orchomenus, see Schliemann, *Orchomenos.* Leipzig, 1881.
  Newton, *Essays on Art and Archaeology.*
  Milchhoefer, *Anfänge der Kunst in Griechenland.*
  Helbig, *Das Homerische Epos.*
  Perrot, *History of Art*, vol. v.
  For latest excavations the recent volumes of the Πρακτικὰ and Ἐφημερίς.

Among the monuments which Athens offers to the traveller's curiosity, among the sights which he must visit after the Propylaea and the Parthenon, the guides point out on the boulevard of the University a large house surrounded by porticoes and surmounted by statues representing the Homeric heroes. It bears on its façade the following inscription, which is at first sight rather surprising and disconcerting: Ἰλίου μέλαθρον—the palace of Troy. Every passer-by will tell you, for it is one of the glories of the town, that this palace is in fact the house of Dr. Schliemann, the great admirer of Homer, the famous archæologist who has explored the ruins of Troy and of Mycenae, and has without doubt discovered the treasure of Priam and the tomb of Agamemnon.* Go further, and seek an introduction to this hospitable mansion, you will be met on every side by the memory of Homer and of the heroes whom he has sung. Nor must it cause you too much surprise, as you are shown the fine collection of antiquities gathered by the master of the house among the ruins of Hissarlik, to hear hundreds of lines of Homer recited even by a woman's lips, and the children of the house called by the harmonious and heroic names of Andromache and of Agamemnon. For in that house Homer is God and Dr. Schliemann is his prophet; and we may well believe that the prophet is more honoured than the god.

I.

Before we begin the story of Dr. Schliemann's marvellous discoveries it will not be amiss briefly to sketch the history of the man himself.

Dr. Schliemann has been his own biographer. At the beginning of the book called "Ilios," in which he related in 1882 the result of his excavations in Troy, he published a long account of his own life and work, a kind of "Confessions" of a singularly attractive

---

* It has been thought best to leave unaltered this notice of Dr. Schliemann, written before his death on December 26th, 1890. For a fuller estimate of his work readers are referred to an article by Professor Gardner in *Macmillan's Magazine*, April 1891.

nature. This autobiography is a very remarkable document, an extraordinary mixture of intentional *naïveté* and frank infatuation, a strange combination, made up in equal proportions of a strongly commercial spirit on the one hand, of a rare talent for business, and profitable business too; and, on the other, of lively religious feeling—a religious feeling, it is true, which partakes slightly of that Teutonic character which is inclined to monopolise the Divine favour for its own benefit. We may trace in it, side by side, a passion for archæology and for science, and a strong infusion of German sentimentality; above all, we find in it that marvellous confidence in himself which it seems to have been the aim of Dr. Schliemann's adventurous life to justify. The life of this man indeed, who, by dint of perseverance only, without means and almost without teaching, succeeded in educating himself and making his own fortune, reads like a romance: and the life of this merchant who spent five-and-twenty years in amassing a fortune, in order to spend a half of his income during the rest of his life in archæological discoveries, is a romance of no common kind. We may forgive Dr. Schliemann much for the sake of the noble use he made of his riches; for the sake of his undertakings, "no less bold than disinterested" (the expression is his own, and we may regret that he did not leave to us the pleasure of suggesting it); for the sake of that thirst for knowledge which was so keen that it sometimes overstepped the bounds of prudence for the sake of that untiring and generous activity which never flagged nor grew weary.

Since, as it seems, there is nothing unimportant in a

great man's childhood, Dr. Schliemann has given us numerous details of his early years, spent in a village of Mecklenburg-Schwerin, and of the "vows of eternal love" which he gravely exchanged, when between nine and twelve years old, with a little neighbour of his own age. This love, we must admit, was more archæological than passionate. By a strange predestination—so remarkable that it may have been, perhaps, to some extent retrospective—Schliemann from his childhood was inspired by a passionate love of Homer and a desire to recover Troy. He talked of it with his little friend, and they agreed that they would be married when they grew up and would go to dig among the ruins of Troy. "Thanks to God," continues Dr. Schliemann, "my firm belief in the existence of that Troy has never forsaken me amid all the vicissitudes of my eventful career. But it was not destined for me to realise till in the autumn of my life —and then without Minna—our sweet dreams of fifty years ago." Minna had not waited, and Dr. Schliemann seems to have consoled himself. At the beginning of the volume devoted to Mycenae, an affectionate preface dedicates the work to Mrs. Schliemann, "as a slight proof of my admiration for her Homeric studies, of my gratitude for her zeal and devotion, and for the energy which sustained my courage when severely tried."

Family misfortunes soon transformed the admirer of Homer and of Minna into a grocer's shop-boy; circumstances afterwards took him as cabin-boy on board a merchant-vessel, and a storm cast him penniless on the coast of Holland. We find him again in Amsterdam, clerk in a merchant's office with a salary of thirty-two pounds a year, lodging for less than seven shillings a month in a fireless garret, breakfasting on a little rye-soup, dining for twopence, and devoting half his income to the completion of his interrupted education. With invincible tenacity, with indomitable energy, whose only drawback is that perhaps Dr. Schliemann speaks too much of it, he devoted his scanty leisure to study, and even encroached upon his hours for meals, in order to learn successively, English in six months, French in as short a time, and then, more rapidly still, Dutch and Spanish, Italian and Portuguese, at the rate of six weeks for each language. Then it was the turn of Russian; and in order to make better progress by reciting aloud all he was learning in that difficult language, he hired a poor Jew for four francs a week to come every evening for two hours and listen, without understanding a word, to Dr. Schliemann repeating his Russian. These studies were to be of great use to him, for it was in Russia that he made all his wealth. At this point the autobiography resembles a ledger: the turnover of Schliemann & Co. is carefully recorded, and the articles of import and export are complacently enumerated.

In eight years Dr. Schliemann made twenty-four thousand pounds; he then doubled this capital in one year, 1854. As a reward he gave himself the pleasure of learning Greek, and he adds modestly that he is perfectly acquainted with all the rules, without even knowing if they are or are not in the grammars. Meanwhile his business was prospering, and he was now making on an average ten thousand a year—a handsome sum for a man who had once been obliged to beg on the quays of the Texel. "Divine Providence," he says, had "marvellously protected him."

He was now rich enough to realise, at last, the dream of his life; and after several long journeys, in which he travelled all round the world from Egypt to the United States by way of China and Japan, he resolved in 1868 to devote to Homer the wealth he had acquired and the remaining years of his life. He visited Ithaca, Mycenae and Troy, and in 1871 he dug the first trench in the hill of Hissarlik. The story of his excavations, frequently interrupted, and six times resumed with unconquerable perseverance, in which he believes himself to have discovered under the superincumbent ruins of five successive settlements, the authentic and indisputable remains of the Troy of Homer, should be read in the book itself. There we find all the characteristic features of the man, the thoroughly commercial love of order which causes him to keep an exact account of the money spent upon the work, the ardent imagination which leads him to recognise the very material of the Homeric poems in the ruins of Hissarlik. Nothing is missing, neither the treasure of Priam nor the jewels of Helen, neither the gates of the city nor the skulls of its inhabitants,—nothing, not even the traces of the great fire which overwhelmed the city at the final catastrophe. Certainly it is an ingenious effort of imagination to discover in the "burnt city," as Dr. Schliemann calls it, the very town of Hector and of Priam, and to reconstruct, down to its minutest details, the plan of the Homeric Ilium; but we must admit that Dr. Schliemann's devotion to Homer is too great for him to be quite impartial in this matter. The recovery of Troy gives a different kind of pleasure, and causes more sensation in the world, than the discovery of some nameless pre-historic town; and Dr. Schliemann's imagination could not resist the temptation. Unfortunately, to any one to whom the profound faith of the explorer is wanting, all these attributions seem somewhat doubtful; but what of that? The tangible results of these excavations, the ornaments discovered, the ten treasures explored, form a sufficiently splendid conquest for science. There are many other things in Dr. Schliemann's autobiography: there is the story of his excavations at Mycenae from 1874 to 1876, and of his explorations in Ithaca in 1878, and at Orchomenus in 1881. But for him these are merely interludes, and though to my mind they are more interesting than the play itself, his thoughts were always turned towards Troy. At the present time (1890), after having explored, with a more truly scientific method, the acropolis of Tiryns, and followed the footsteps of Homeric heroes at Pylus and in Laconia, we are informed that Dr. Schliemann is returning to Hissarlik, and we must do outspoken homage to the energy and generous activity of a man to whom archæology is indebted for discoveries of the utmost importance.

II.

Such is the man. Now let us turn to his work as it appears in one of its most brilliant episodes—the exploration of the acropolis of Mycenae.

There are few people, I imagine, who have not heard the name of Agamemnon: if they do not know him from Homer, they will at least have met with him in Browning or certainly in Lewis Morris. It is not so generally known that the king of men reigned, according to tradition, in a town of Argolis called Mycenae, and there is still greater ignorance of the fact that considerable ruins of this town still exist. At the farther end of the plain of Argos, near the little village of Kharvati, and precisely at the opening of the narrow and picturesque pass through which the road runs from Argos to Corinth, on an extensive rocky plateau triangular in shape, there

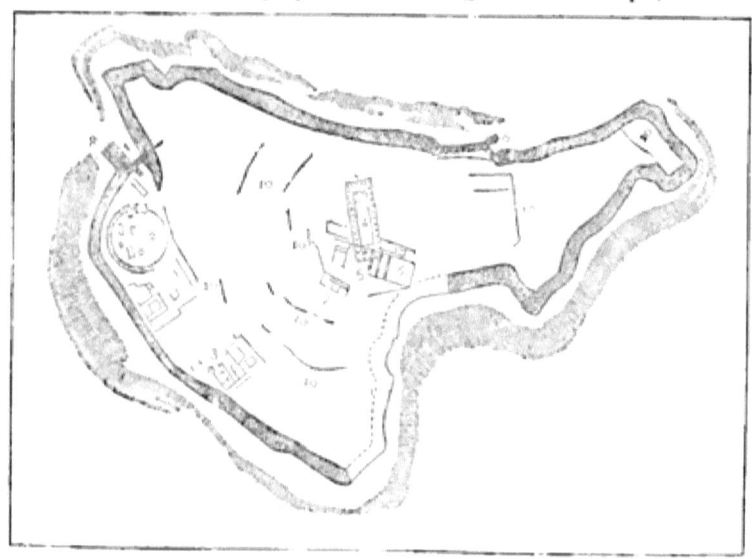

Plan of Mycenae.

1 Lion Gate.
2 Circular Enclosure of Graves.
3 Pre-historic House.
4 Doric Temple.
5 Court of Palace.
6 Megaron.
7 Staircase.
8 Towers.
9 Postern.
10 Supporting Walls.

rise the ruins of the ancient citadel of the princes of Mycenae. On one side a deep ravine defends the approach to the fortress; the whole circuit of the acropolis is surrounded by lofty walls whose height varies from thirteen to nearly thirty-three feet, and which are more than sixteen feet thick. The way in which these walls

The Lion Gate.

are built is worthy of notice; the blocks are not laid in regular courses and carefully cemented; polygonal stones built up without mortar, and, in the parts which are most carefully worked, quadrangular blocks laid in horizontal courses but not bound together by any mortar, make up these fortifications. The name Cyclopean is given to this kind of building. A second circuit in the interior, nearly thirty feet high, completes the system of defence.

The fortress has two entrances—a little postern on the northern side, which may have served for unexpected sallies in the rear of a besieging force, and the principal gateway at the north-west angle of the citadel. Between the circuit-wall and a large quadrangular tower commanding the approach, there extends a roadway nearly fifty feet long by rather less than thirty wide, at one end of which rises the famous Lion Gate. It is an almost square opening, formed by two uprights on which rests an enormous stone lintel. Above this is a triangular space closed by a slab of basalt, on which are carved two lions heraldically opposed, whose paws rest upon an altar surmounted by a column; it is well known as one of the earliest pieces of Greek sculpture. Thus, as it were, the sovereigns of Mycenae set up their scutcheon at the entrance of the acropolis, and by its design, essentially Asiatic in origin, took pleasure in recalling their Oriental descent.

Below the citadel extended the lower town, also surrounded by fortifications. There too we meet with remains of Cyclopean buildings, the most remarkable of which is the treasury of Atreus. A "dromos" or passage, 100 ft. long by 20 ft. broad, lined by rather high walls, leads to a doorway, surmounted by an enormous stone lintel, through which we enter a large circular chamber about 50 feet high, with a dome-shaped roof. The walls were formerly, in accordance with a custom mentioned in Homer (*Od.* vii., 84—87), covered with plates of metal; bronze ornaments fastened to the doorway, and a coating of coloured marbles, decorated the exterior of the monument. In this splendid building, which was doubtless merely a tomb, popular credulity early insisted upon discerning a royal treasury; and the report of the riches which were believed to be buried there has more than once attracted the greed of treasure-seekers. The treasury of Atreus was stripped by a Turkish governor, Veli-pacha, of the ornaments which covered it,[*] and it has come down to us empty. The same fate has befallen the other buildings of the same kind, six in number, discovered in what remains of the lower town, one of which, partially excavated by Mrs. Schliemann, wife of the archæologist, has received her name.

It will no doubt be asked how buildings of such great antiquity, the remains of a prehistoric period, have been preserved almost unharmed, while so many cities still flourishing in classical times have disappeared, leaving scarcely a trace behind them. This surprising duration is due to the fact that Mycenae early ceased to exist as a city. In the beginning of the fifth century the jealousy of Argos proved fatal to its ancient rival, which had already lost much of its early splendour, and Mycenae fell in spite of the desperate resistance offered by its inhabitants behind the walls of the citadel of

---

[*] Some fragments are now in the British Museum.

Agamemnon. Undoubtedly a village existed upon the same site down to the end of the second century, but from the time of the Romans the place was deserted. The Slavonic invaders, the men of the Middle Ages, Byzantine and Latin, and the Turks, passed by these desolate ruins without pausing ; no great neighbouring town sought materials there for its buildings, and the citadel of the Atridae, consecrated by the great memories of the Homeric poems and by the tragic legends which clustered around its walls, still rose upon its solitary rock.

Few towns, indeed, enjoyed in early ages a more brilliant fame. History and poetry vie with one another in bearing witness to the might and the riches of the sovereigns of Mycenae. Homer calls Mycenae "the wide-wayed"* city, "rich in gold" ($πολύχρυσος$ †). Thucydides, too, mentions its reputation for wealth.‡ Legendary history assigned its origin to Pelops, the Phrygian, from the land of Pactolus, the remote founder of the dynasty of the Atridae, and told how Atreus and Agamemnon had extended their empire over a great part of the Peloponnesus. The tragic history of the house of Pelops gave no less interest to the walls of Mycenae : Agamemnon massacred on his return from Troy by Clytemnestra and Aegisthus ; Orestes returning afterwards to avenge his father's death by the murder of his mother,—all these subjects, which grew famous in the hands of the tragic poets, had for their background the acropolis of Mycenae. This was the theatre in which

---

* *Iliad* iv. 52 : εὐρυάγυια Μυκήνη.
† *Iliad* vii. 180, *Od.* iii. 305 : πολύχρυσος.
‡ Thuc. i. 9.

were played the early scenes of the *Orestes* of Euripides and the *Electra* of Sophocles, and it would seem, from the exactitude and precision with which the poet describes the citadel of Mycenae and the surrounding country, that he had himself visited the scene of the tragic events which he put upon the stage. So many famous memories gave an especial interest to the ruins of Mycenae. From the earliest times its visitors were numerous; and just as in Jerusalem to-day one tries to connect some special memory with each stone, so at Mycenae they tried to discover the precise spot where so many dramatic incidents had taken place. The guides not only pointed out the Lion Gate, the circuit wall, and the treasury of Atreus, but they asserted that they could show the sepulchres of the kings of Mycenae, and especially the tomb of the illustrious victims of the Trojan tragedy. A traveller of the second century [*] relates how he was shown, on his visit to Mycenae, the five tombs of Agamemnon and of his companions, and also the graves of Clytemnestra and of Aegisthus, who were buried at no great distance, but outside the sacred precincts within which their victims rested. It is to this statement of Pausanias that we owe the discoveries of Dr. Schliemann.

To find the tomb of Agamemnon—what a dream for an archæologist! what a temptation for such a passionate lover of Homer as Dr. Schliemann, his appetite already whetted by the discovery of Troy! As a matter of fact, all those who had up to that time studied the words of Pausanias placed the tombs of which he speaks in the lower city; but Dr. Schliemann was not a man to be alarmed or disconcerted by contradiction. He main-

[*] Pausanias, ii. 16

tained energetically that these famous monuments must be sought on the acropolis itself.

In February, 1874, he began the work by sinking trial shafts. Two years later, on August 7th, 1876, he definitely hazarded the venture. He must certainly have felt a momentary alarm as he surveyed the vast acropolis covered with ruins, and asked himself anxiously where he should begin. But he had faith. He wished to find

Ancient Tombstone.

the tomb of Agamemnon; he was convinced that he would find it. He had money, too, which, even more than faith, moves mountains; and above all he had the indomitable tenacity which is not discouraged by any check. Success has crowned his efforts by discoveries which are perhaps the most remarkable of our time.

On August 7th, 1876, Dr. Schliemann ordered the workmen to begin to dig near the Lion Gate, and after having brought to light the ancient threshold, he found himself upon an extensive terrace covering the western side of the hill; here excavations were actively pushed on. Soon, at a distance of a few yards from the surface, they found pieces of archaic pottery and remains of Cyclopean buildings; and among the *débris* they brought to light were nine large slabs of stone, or stelae, in which Dr. Schliemann immediately recognised the sepulchral monuments of the royal necropolis which he was seeking. Among these stelae, four were decorated with rude sculpture, childish and barbarous in style—a sure sign of the high antiquity of the objects he was about to discover. The excavations which followed soon showed that these tombstones had been erected in the midst of a circular enclosure about 82 feet in diameter, surrounded by a double circle of stones fixed in the ground, upon which other slabs were laid horizontally in such a way as to form a kind of bench or seat round the open space.

Dr. Schliemann at once thought he recognised the place. He remembered that in Homer [*] the old men sit in a circle on polished stones within the sacred enclosure; he remembered, too, that these public places where the king's counsellors met to deliberate were formed of large stones sunk in the soil and skilfully adjusted, and that Euripides, in his tragedy of *Orestes*,[†] speaks of a circular agora at Mycenae; lastly, he remembered that in antiquity citizens of distinction were buried without hesitation in the midst of the place of assembly.[‡] From this moment doubt was no longer possible: this circular enclosure, situated at the very

---

[*] *Iliad* ii. 53-96-99; *Od.* viii. 4-7.
[†] *Orestes* 919: ἀγορᾶς κύκλον.
[‡] Pausanias, i. 43. 3.

entrance of the acropolis of Mycenae, was the agora of the city. The remains of Cyclopean work which surround the enclosure formed part of the palace of the Atridae, and as to the stelae they proved that the tombs so ardently desired were now at hand. It only remained to find them.

Fortune aided Dr. Schliemann beyond all hope. At the end of October, 1876, below the spot where the stelae had been found, there was discovered a quadrangular cavity 21 ft. 5 in. long by 10 ft. 4 in. broad, hewn in the side of the rock. It was undoubtedly a tomb. Unfortunately it was almost empty; and for a moment it was feared that the graves they were seeking had been broken open in antiquity. Soon, however, there appeared a second tomb at a depth of 15 ft. below the level of the rock; and here they discovered, stretched on a layer of pebbles, three skeletons whose heads were turned to the east. Three other similar cavities were successively discovered within the circuit of the agora, and shortly afterwards a sixth was found outside it. Each of these tombs contained several bodies; the bones of seventeen persons in all were found there, three of whom were women and three children; and one of the corpses, thanks to a kind of embalming which it had undergone, was almost intact. All the bodies, moreover, seemed to have been placed in the tomb in a singular way: instead of laying them lengthwise in the grave, they had been placed parallel to its shorter sides; and as the cavity measured scarcely 5 ft. 6 in. in breadth, it had been necessary in more than one instance to press the corpse down with considerable force into a space too narrow for it. Upon this circumstance

— which was certainly remarkable, but which no doubt arose from the desire to give for religious reasons a particular orientation to the dead body—Dr. Schliemann built up an ingenious romance, to which we shall return later. We must first finish the story of the discovery, and point out in what its exceptional interest consists: in the quantity, that is, of ornaments and articles of value accumulated in these graves.

It was a very ancient custom among the nations of antiquity to bury with the dead the objects they had used or cherished during their life on earth. It was the habit of the Egyptians, and the excavations of Tanagra will soon show us how long this tradition endured. The same custom existed in primitive Greece. Not only was the dead man arrayed in his richest garments, with his arms and jewels by his side, but care was even taken, if he were a man of rank, to secure that he should have in death a following worthy of his dignity.

We see Achilles in the *Iliad* sacrificing prisoners upon the tomb of Patroclus,* to do him service in the lower world; and it is probable that a similar reason will explain the great number of corpses buried in the graves of Mycenae.

The incomparable splendour with which these dead are buried has never been equalled elsewhere. They were dressed in robes of state to leave this world, and were laid in the tomb with their golden diadems upon their foreheads, wearing their richest jewels and most sumptuous raiment. On their heads were crowns, belts and baldrics of gold around their breasts, while their faces,

---

* *Iliad* xxiii. 171-84.

by a singular custom which seems to have been borrowed from the East, were covered by a golden mask which reproduces the features of the deceased. Their garments were ornamented with thin plates of gold, seven hundred of which were found in a single tomb; and their richly-inlaid weapons, whose sheaths are curiously adorned with bosses of gold, lay within reach. The women were no less splendidly attired: they also wore diadems on their heads, necklaces round their throats, rings on their fingers, as well as brooches, earrings, and bracelets of admirable workmanship. Lastly, by the side of each of the corpses were placed vessels, often of gold and silver, which contained the provisions needful for the sustenance of the dead in the lower world. Certainly these were no insignificant personages buried here,—they were sovereigns interred in state, and in this respect at least Dr. Schliemann's hopes were not deceived.

It is impossible not to recognise the extreme importance of these memorable excavations, and the surprising harmony which exists between their results and the legends of the wealth of Mycenae. Estimated simply at the value of the gold which they contain, the jewels discovered in the tombs of the acropolis are worth more than four thousand pounds; while their value from an artistic or scientific standpoint is simply incalculable. They reveal to us in fact with positive certainty the civilisation and social state of Greece three thousand years ago; they reveal the existence, long before the Dorian invasion of the Peloponnesus, of a rich and powerful empire, to which tradition attaches the name of the Pelopidae; they reveal the influences under which society in the heroic age attained its development, and the primitive art of Greece arose. These are splendid results, and sufficient to reward more than one archæologist. Nevertheless, Dr. Schliemann was not content. He wished to deprive these dead men of their incognito, and to give a name to these masks and corpses: as to the name itself he could not hesitate. It was to discover the tomb of Agamemnon that he had set to work, nor did he doubt for a moment that he had succeeded in doing so. On November 28th, 1876, he despatched the following triumphant telegram to the King of Greece: " I have the greatest pleasure in announcing to your Majesty that I have discovered the tombs which tradition, according to Pausanias, pointed out as the graves of Agamemnon, Cassandra, Eurymedon, and their companions, all murdered at a banquet by Clytemnestra and her lover Aegisthus. I have found immense treasures in these graves, sufficient by themselves alone to fill a great museum, which will be the most wonderful in the world, and for centuries to come will bring thousands of strangers to Greece. God grant that these treasures may become the foundation of great national riches!" At the same time Dr. Schliemann proclaimed to the four winds that he had found the actual corpse of Agamemnon: this face with its golden mask, on which is still imprinted the majestic smile of the king of men, could be his alone; and his, too, the corpse with eyes half closed and jaws scarcely meeting—sure signs of the criminal negligence with which Clytemnestra forgot to pay the last respect due to her murdered husband. This seductive hypothesis explains every detail of the discovery and the ingenious romance constructed by Dr. Schliemann, and which Mr. Gladstone has even surpassed in his preface to the volume called "Mycenae," deserves to be read. If the remains of Agamemnon, in spite of his ignominious death, have been given honourable burial in the agora of Mycenae, it was because the murderers, fearing a popular outbreak, dared not refuse the last rites to the king of men, but this satisfaction once ostensibly given to the sympathies of the multitude and the grave made ready, the corpses were hastily thrown in, careless of the hurt they might receive through this unseemly desecration. It was not until later years when Orestes had taken startling vengeance on the murderers, that he reopened the grave of the conqueror of Troy to render him the last honours. Then solemn reparation was made to the dead : they were arrayed in their burial robes, and the bodies burned in accordance with sacred custom ; and finally, in order to point out to popular reverence the spot where these illustrious dead were resting, the restored government set up in the midst of the agora the funeral stelae which marked the situation of the royal burying-ground.

Certainly all this is extremely ingenious and subtle, —perhaps too ingenious and too subtle to be true. If we suppose that Agamemnon ever existed outside of the imagination of poets, which is by no means proved, upon what evidence do we presume to connect the graves of Mycenae with his name ? There can be no doubt that the tombs opened by Dr. Schliemann are in reality the graves of kings—obscure private individuals were not buried with such splendour. But, admitting this, what must we think of the rest?

Here is Pausanias, replies Dr. Schliemann, whose account plainly proves that it is a question of Agamemnon. Unfortunately the author's text is not as certain as is necessary for the argument; it is susceptible of more than one rendering, and we may quite as easily conclude from it that these illustrious graves are to be sought in the lower city. Dr. Schliemann himself allowed that the graves discovered on the acropolis were probably no longer visible in the time of Pausanias, buried as they were under the fallen earth from the upper part of the citadel. Finally Pausanias expressly names five tombs; and Dr. Schliemann's excavations came to an end at the precise moment when he had discovered within the sacred circle the mystic number of graves necessary for his demonstration. Unfortunately the Greek government, attracted by such splendid discoveries, continued the work on its own account, and a sixth tomb was found no less rich than the former. This additional tomb is in danger of being a tomb too many for the hypothesis. It is certainly more attractive to accept Dr. Schliemann's romance without discussion, and to believe that we possess the mortal remains of Agamemnon, just as we have recently discovered in Egypt the mummified corpse of Rameses II. but the great Pharaoh had his documents with him in his tomb—Agamemnon has none. After all, what does it matter to us? The royal tombs of Mycenae, whatever were the name of the princes buried there, have revealed to us a whole unknown period of Hellenic civilisation; and this is of more value than the discovery of the fabulous remains of Agamemnon.

Hitherto no writer would have dared to trace the

history of Greece further back than the eighth or ninth century, or the history of its art beyond the sixth or seventh. To-day, thanks to the excavations of Mycenae, we can ascend much further: we can connect the early ages of Greece with the ancient Oriental civilisations of Egypt and Chaldea, of Assyria and of Phoenicia, by whose side, to use the words of Plato, the Greeks were but children; we can reach those remote ages beyond the horizon of Homeric poetry, which are separated from us by more than three thousand years; and thus bringing into the region of history centuries which were only known to us by religious or poetical traditions, we can, thanks to our definite information, carry into those distant ages the clearness and the certainty of scientific method. This is worth more than merely placing this civilisation of a past age under the name of Agamemnon, a convenient and simple method which answers to everything but teaches nothing. Now, there are many things to learn in the civilisation of Mycenae; it is not enough to affirm its existence, it must be explained as well.

III.

The telegram in which Dr. Schliemann announced his discovery threw the learned world into a state of extreme excitement. The study of the objects found in the course of the excavations still further increased this surprise, for it was an entirely unknown world which they revealed. The ordinary types of Greek art were sought in vain. No familiar face, if I may say so, was found among them. Archæologists were struck by the strange and barbarous character of this civilisation, and by the heaviness and intricacy of its art; the designs they found were foreign to the accustomed style of Greek art, while on the other hand they presented, in place of the naïve clumsiness of primitive work, such a heavy and overloaded elegance, that many competent judges assigned them, under the influence of this first impression, to a civilisation in its decadence rather than to an early stage of art. In the face of Schliemann's romance other romances arose, equally ingenious and equally false. One eminent archæologist unhesitatingly assigned these ornaments to the Byzantines of the Middle Ages, and declared that they dated from the twelfth century of our era, almost in the same way that the officers of the expedition to the Morea were deceived by the heraldic aspect of the lions sculptured on the gate of Mycenae, and saw in them a work of the Middle Ages. Others, deceived by the points of resemblance between the objects discovered and the ornaments of the barbarous tribes of the North, and struck by the analogy which existed between these monuments and the tombs of the Crimea, imagined that the graves discovered at Mycenae were those of some barbaric chiefs, either of the Gauls who ravaged Greece in the time of Pyrrhus, or of the Heruli who invaded the empire in the fifth and sixth centuries; and they thought that in these graves were heaped up at random the products of the workmanship of the north, and Greek ornaments stolen by the barbarians from the temples of Argolis. Others have constructed still bolder romances; and upon the evidence of imaginary resemblances discovered between the ornaments of Mycenae and the objects found in the necropolis of Hallstadt in Germany, or in those of Hungary, Denmark, and Sweden, they have imagined a race, unknown to history, coming from the North, which between the period of the Dorian invasion and the sixth century, established itself upon this rock of Mycenae, and lived there without intermingling with the neighbouring Greek tribes.

To-day, in spite of certain repetitions of the attack, all these disputes are at an end. Numerous other discoveries have brought to light, not only upon the soil of Argolis, but in various other localities in Greece, at Spata, at Menidi in Attica, even upon the acropolis of Athens, and from Boeotia and Thessaly in the north to Laconia in the south, objects of the same kind and of the same style as those discovered at Mycenae. Resemblances at first unperceived have been established between these and other objects found in the islands — at Santorin and Rhodes, and quite recently in Crete and in Cyprus—and thus these excavations, extending beyond the narrow limits of the plain of Argos, have proved at once the antiquity of the objects discovered and the extent of the civilisation to which they bear witness. Lastly, new and important excavations have been made quite recently at Mycenae itself, and have added new elements to the delicate and difficult study of this vanished civilisation.

Dr. Schliemann, after his fortunate campaign of 1876, had not left Mycenae without meaning to return. It remained to remove the *débris* accumulated upon the summit of the acropolis, and to explore the numerous groups of tombs scattered around the lower city. The Archæological Society of Athens, by taking upon itself this task, did not leave Dr. Schliemann the glory of completing his work. The Society began its excavations in 1886 on the summit of the citadel, and had the good fortune to discover beneath the ruins of a Doric temple the *débris* of a large palace, in which the royal abode of the sovereigns of Mycenae has been gladly recognised, and which seems to have been decorated with very curious paintings, some fragments of which have been found. Thanks to this building, which is contemporary with the tombs discovered by Dr. Schliemann, and the arrangement of which recalls the palaces described in the Homeric poems, another aspect of this Mycenean civilisation is unfolded before our eyes, whose especial interest we shall explain when studying the excavations of Tiryns. At the same time the explorations made at the foot of the acropolis and on the neighbouring slopes were not less successful; in 1887 and 1888 fifty-two tombs were explored, and a great number of objects in gold and silver were found, as well as in bronze and ivory, together with pieces of pottery, very curious engraved stones, and—which is remarkable, for Mycenae had hitherto given no proof of the use of the metal—two iron rings. All these discoveries presented striking analogies with those made before on the acropolis; and even the arrangement of the tombs offered no fewer points of resemblance. Arranged around the lower town in a certain number of little groups, several of which have as their centre one of the domed tombs reserved for people of rank, these graves, almost all of which contain several corpses, are entered by a rather long passage-way, filled up in order to close the entrance to the tomb; and the dead who were laid there are buried in the same way as the bodies found within the sacred circle of the fortress.

These discoveries, which have taken place quite recently, have again drawn attention to this Mycenean civilisation; and without pretending, in the present state of our knowledge, to offer a full solution of an enigma which is perhaps to remain for ever unsolved, we cannot escape certain pressing inquiries raised by this difficult problem. To what race does this brilliant civilisation belong, the traces of which we find upon the acropolis of Mycenae? In what century did it arise? under what influences was it developed? In what respects was it original? Is it, lastly, an isolated phenomenon in history or a part of a greater whole? These are questions to which learning has offered, and still offers, contradictory answers, and of which we must only seek solutions of the most general kind. In an enquiry which is continually renewed, and to which every day brings new materials, it would be imprudent to press the truth too closely, and, in the study of this complex civilisation, wisdom consists in distinguishing, in the first place, the different elements of which it is made up.

If the objects discovered in the course of the earlier or the later excavations at Mycenae are studied from the standpoint of merit of style or skill in workmanship, even a superficial observer may easily separate the jewels and weapons and pottery and paintings into two classes. The workmanship of some is still coarse and barbarous; and although we find in them already the signs of a natural genius for the arts, this genius is nevertheless only just awakening, and is still making its first attempts. Others show more advanced technique and a style which is more completely formed. There are two distinct civilisations represented here; the one is still rudimentary, while the other has already reached a degree of refinement which infers a long previous development. We must ascertain, if possible, the precise limits of these two classes, mingled in the same grave; must find whence comes the more highly-developed art whose presence is undeniable, to however small proportions some may have wished to reduce it; and must discover what are the distinctive and original characteristics of the indigenous art.

The historical importance of the legends which we find current in the childhood of a nation is well known. Primitive peoples like to express the ideas which have guided them in a concrete form, as long series of events modifying their destinies; they like to personify these gradual and obscure influences under a few great names, and in this way their legends throw light upon their history. It is by no means unimportant, then, that we should meet in the heroic traditions of Argolis with foreign princes, who bring with them great advances in civilisation. There are Danaus from Egypt, and Pelops from Asia; and we find foreign heroes landing in the same way on other parts of the eastern coasts of Greece, the Egyptian Cecrops in Attica, and the Phoenician Cadmus in Boeotia. Translated into plain prose, these poetical legends mean that the imagination of the Greeks has summed up in these famous expeditions a thousand unknown journeys which gradually brought into Greece, from the coasts of Asia, the arts of the great civilisations of the East.

Let us turn now to the historians and poets. They, too, have preserved the memory of great maritime powers, which from the coasts of Asia transmitted, together with their merchandise, the knowledge of the arts to the islands and to Greece. There is the great Phrygian empire established in the defiles of Sipylus and round the mouth of the Hermus in the twelfth century before our era, or even earlier, with which the fabulous names of Tantalus and Pelops are connected, and whose monuments offer more than one striking analogy with those of Argolis. There is a Carian Empire,* the date of which is lost in the darkness, but of which the memory survives throughout the Aegean, whose sovereignty extended over the islands, and even to Argolis, at the time when these bold adventurers, sailors, merchants, and pirates all at once, were establishing the first permanent relations between the nations in the eastern basin of the Mediterranean. There were the Phoenician navigators of the fourteenth and thirteenth centuries before our era, whose commercial activity made them the intermediaries between Greece, still uncivilised and barbarous, and the countries of the East.

Even in Homeric times, all objects of value come from Phoenician hands. On every page the poet extols Sidon, "rich in bronze," and praises the beautiful embroideries, the rich jewels, the marvellous weapons, the splendid ornaments, which the merchants of Phoenicia offer to the admiration of Greece. These bold and skilful sailors had already long been pushing their expeditions and their factories further, day by day. Every year they came to sell the treasures accumulated in

* Herod. i. 171; Thuc. i. 4.

the warehouses of Tyre and Sidon to the Greek tribes, whose fresh and youthful minds were awakened by them to a feeling for art. Unscrupulous, moreover, very cunning, very skilful in deceiving, and over-reaching, merchants and pirates at the same time, they sold and stole by turns. They came to Greece to exchange for the produce of the country the articles of luxury which they brought from their workshops, brilliant stuffs figured or dyed with purple, splendid embroideries, silver vessels artistically chased, wonderful ornaments; but they did not disdain upon occasion to act as slave-dealers, and to carry off from the coasts of Greece young men and maidens for whom they gained high prices in the markets of the East. In the beginning of his history,[*] Herodotus shows us Phoenician merchants, the commercial travellers of antiquity, bringing the wares of Egypt and Assyria into Argolis, producing jewels and rich stuffs for the admiration of the women, and availing themselves of these attractions to allure the young girls of Argos on board their swift ships, and thus carry them off. The Odyssey is full of similar adventures, of which the best known is that of Eumaeus, the faithful servant of Ulysses.[†]

The father of Eumaeus was king over two cities in one of the islands of the Archipelago. One day, Phoenician merchants were seen to arrive in the country —"mariners renowned, greedy merchantmen," who brought in their black vessels the goods which they were accustomed to exchange with the Greeks. Now, the king of the country had among his slaves a woman of Sidon, "tall, fair, and skilled in bright handiwork."

[*] Herod. i. 1.   [†] *Odyssey* xv. 414.

As she was washing on the seashore she made the acquaintance of her countrymen, and one of them spoke to her sweet words of love; "for love," says the old poet, "beguiles the minds of womankind, even of the upright." The strangers proposed to the slave to take her back to her native land; and in order to facilitate her escape, they feigned not to have noticed her, and began to sell their goods. On the day of their departure one of them went to the king's house carrying one of their necklaces of amber and gold, such as have been found in the excavations of Mycenae; and whilst the women of the house crowded round to admire the jewel, the Phoenician made a sign to the slave. She left the palace, carrying away three cups, and the king's young son; and a few hours after the strangers had disappeared, going to Ithaca, where they sold Eumaeus as a slave.

Such is the artless story we are told. It shows the relations which existed between the Phoenician merchants and Greece in the Homeric age. Things were much the same some centuries earlier, when Mycenae the Golden reigned over Argolis.

This is not all. The Egyptian documents which record the victories of Seti I., Rameses II., and Menephtah, show us the "nations of the sea," Trojans, Mysians and Lycians, Pelasgi, Tyrrhenians, Achaeans, and Schardani, in the fourteenth century before Christ, twice allying themselves with the populations of Syria and Phoenicia in order to attack the monarchy of the Pharaohs. Nothing proves more clearly the antiquity of the relations which united the tribes of Asia Minor, of the Cyclades and of Greece to the Phoenicians; and we catch a glimpse of incessant migrations in those distant centuries across the eastern basin of the Mediterranean, of constant relations between the two coasts, and of an undoubted influence exercised by the more advanced nations on the arts and the manners of the less civilised. Now, what legend suffered us to divine, what history permitted us to conjecture, archæology has undertaken to fully prove, by showing the place held by objects imported from the East in the civilisation revealed by the excavations of Mycenae.

It is from the East that those vessels of gold and silver come, whose graceful forms, delicate ornamentation, and skilful technique would suffice to attest their foreign origin, even if a singular proof were not forthcoming to complete the demonstration. In the paintings of an Egyptian vaulted chamber at Thebes, we see the tributary nations bringing their presents to the Pharaoh, Thothmes III.; and among them the inhabitants of Kefta—who are Phoenicians—present vases whose shapes strikingly recall those of Mycenae. They offer —and here the analogy is still more close—heads of oxen in silver, enriched with golden horns, which are exactly similar to a monument of this kind found at Mycenae, and which from an artistic point of view is very noteworthy. From the East come those fantastic animals of gold and ivory, Egyptian sphinxes and Assyrian griffins, and those lions rampant on the vases, whose excessive and altogether conventional length betrays their foreign origin. That fine lion, too, cast in solid gold and chased with care, whose very aspect reveals Egyptian work, comes from the East; and thence come the ivory trinkets, the figurines of women with foreign faces, the mitred heads redolent of Assyria, the ostrich eggs adorned with patterns, the fragments of Egyptian porcelain, the purely Asiatic ornaments, and the animals face to face or locked in mortal struggle. To prove our point more clearly, let us examine those jewels which differ so profoundly from the art of Mycenae—the seals and the skilfully wrought perforated ornaments of chased gold belonging to a necklace, the subject of which—lion-hunting—is frequently found in the monuments of Egypt and Assyria, and the highly characteristic style of which displays a mixture of Egyptian and of Syro-Chaldaean influence, which of itself is sufficient to show Phoenician handiwork. Other proofs may be adduced. A large seal, which recalls Babylonian cylinders, bears engraved upon it, besides three women in characteristic dress, the symbols of the Chaldaean cosmogony—the lynx's head which represents the planets, and the celestial stream which is the Milky Way. These figures of goddesses with doves upon their heads, this little golden temple, with doves again upon the corners, mark the cultus of the great goddess of Babylonia and Phoenicia, the powerful Istar. Above all, these magnificent daggers,* the blades of which are ornamented

Sword-blade.

* *Bull. Corr. Hell.*, x., p. 341.

with an engraved plate of gold or with a bronze plate inlaid with gold of different colours, by their skilful technique as well as by the subjects they represent—panthers pursuing aquatic animals, scantily-clothed men hunting the lion—by the length of the bodies and the character of the design, forcibly recall Egyptian work, and have no local characteristics.

But among these objects which recall Egypt and Chaldaea, and some of which, no doubt, came directly from those distant countries, have all the same origin? Even in the daggers of which we have just spoken, we find an imitation of Egypt rather than actual Egyptian workmanship. It is, as M. Perrot says, art Egyptianised, rather than Egyptian art. Now, amongst all the nations, which is it whose art has been able to imitate the productions of Egypt and Assyria, to mingle these two dissimilar elements in its own artistic creations, and to diffuse its products throughout all the Mediterranean? It is Phoenicia. The great empires of Egypt and Assyria, powerful as they were, had no navy; it was the seafaring people of Tyre and Sidon whose mission it was to make the arts of these countries known in Greece. Their imitative genius first underwent Egyptian influence; a vassal of Egypt until the twelfth century, Phoenicia was long a stranger to Assyria; and it is on that account that we find so few Assyrian objects at Mycenae. It was not till later, in the tenth and ninth centuries, that the Phoenicians brought to Greece the models they had borrowed from the great Mesopotamian empire. This is the second period of Phoenician influence in Greece. But by the side of these foreign importations a national art reveals itself,

Native goldsmiths existed at Mycenae, who were already skilful, and possessed of some of the secrets of the craft. From their hands came the golden masks which covered the faces of the dead, and which all the evidence leads us to believe were made on the spot, as well as those thin plates of stamped gold, the method of whose manufacture has been disclosed by the excavations. Hollow moulds have been found in granite and in basalt, presenting some of the designs which the gold plates offer in relief. The goldsmith did not fill them with molten metal; but with the aid of a hammer he beat into the mould the gold leaf which he wished to stamp. Certainly the Mycenaean gold-beaters may sometimes in their productions have been inspired by Asiatic designs, and may have rudely imitated those monuments of a superior civilisation; but the fact of local manufacture remains. We may assign to this local manufacture the common pottery found in great quantity in the ruins of Mycenae, and the paintings which decorate the palace of the kings; in all of these a particular style makes its appearance —a national genius, of which we must determine the original characteristics.

The designs associated with this decorative style seem at first sight singularly complicated; in fact, they may be reduced to a rather small number of very simple elements, which may be traced back to three principal types. Sometimes they are geometrical ornaments, not formed, however, by the use of straight lines which multiply combinations by intersections at right angles; they are curves, spirals, scrolls, circles, which are worked out with freedom and variety. More often the artists seek their models in nature with a careful observation and a delicacy of visual impression which are already remarkable. It is vegetable life, with its flowers, its leaves, and especially its aquatic plants it is animal life, with its flies and its butterflies, and especially all the fauna of the sea and the shore—shells of every kind, the octopus, the jelly-fish, the starfish —which inspire the goldsmiths and the potters of Mycenae. Soon too, this art, at first so childish, extends the field of its studies; it attempts the representations of animals, especially of dogs and horses; nay, more, it ventures to depict the human figure processions of warriors appear in the pottery and in the paintings, whilst on one fine silver vase recently discovered, the goldsmith avails himself of the resources of gold and enamel.* In these latter developments Mycenaean civilisation already approached the style of those ancient vases found at Athens—that Dipylon civilisation, as it is called, which seems to have succeeded in Greece to the culture of which Mycenae is the type. If our attention is directed, among these various elements, to the noteworthy position given to marine fauna and flora, we shall conclude that the choice of these subjects could only have been made by a people of fishers and sailors. This people had already reached a high degree of riches and power; it was possessed of considerable skill in the industrial arts, and of a flourishing civilisation. It could itself manufacture, and sometimes with a sureness of hand which bears witness to long practice, objects original in style. Its maritime and commercial relations, on

* Mitth., viii. 1.

the other hand, were extensive, for it cannot have found on its own soil the gold and silver which its workmen employ so liberally. Lastly, it had undergone an Asiatic influence, of which the Phoenicians were the medium, and which was principally Egyptian. But when all this is said, what was this people? Are these princes, who sleep in the tombs of Mycenae, the ancestors of the Greeks of history? Are they really those Achaeans whose memory is immortalised in poetry and legend? or are they strangers? Do they belong to some of those races whose maritime supremacy in the Cyclades is recorded in history? Did they come from Phrygia, as so many of the designs found at Mycenae would make us believe—from the lions face to face on the gateway of the acropolis, to those beehive graves of which Phrygian art offered the first model? Do they belong, as some have maintained, not without probability, to that Carian race which was formerly supreme in the Archipelago, and extended its colonies as far as Hermione and Epidaurus? The manner in which the dead are buried at Mycenae, the resemblances between their tombs and those recently discovered in Caria, the analogy between the methods of the Mycenean gold-beaters and the technique of the Carian hammered ornaments, might serve to sustain this hypothesis. Or are they Cretans rather, as a recent theory affirms? and do they date from the time of that first Hellenic empire founded in the Archipelago, to which tradition attached the name of Minos? It is very difficult to solve such problems, very difficult too to give a date to this civilisation. It is allowable to hesitate between those who put its date back to the seventeenth century before the Christian era, and those who would bring it up to the eighth century B.C.; and on this point we must perhaps adhere to those who take a middle course, and for many reasons place the civilisation of Mycenae in the twelfth century before our era.* As to the rest, as to the origin of those princes whose age-long sleep Schliemann has disturbed, perhaps we shall never know anything; perhaps, as M. Perrot has happily said, this nation of builders and hoarders will remain masked in history as were the faces of its sovereigns in the tomb.

IV.

But in default of more precise statements, history may make some conquests at least in those remote centuries, and may succeed in stating some more general conclusions.

The civilisation of Mycenae is not in fact an isolated phenomenon in the history of the early ages of Greece; other discoveries have enabled us to compare analogous objects with the Mycenaean monuments, and have proved that the state of civilisation which we find at Mycenae was at one time that of all the eastern basin of the Mediterranean. In 1877 the excavations carried on at Spata in Attica displayed, in the tombs explored there, the same medley of native and Asiatic productions, and revealed by the side of Oriental importations still more numerous than those of Mycenae, and beside a foreign influence still more extensive and distinct, incontestable

---

* Cf. Petrie, *Egyptian Bases of Greek History*, J. H. S., xi. 271. See also xii. 199.

traces of a local manufacture which presents the same characteristics as those of Mycenaean art. In the beehive tombs of the Heraeum near Argos, and of Menidi in Attica, in the monuments of the same kind explored in 1881 by Schliemann at Orchomenus, and recently excavated at Dimini in Thessaly (1887), and at Vaphio in Laconia, the plan is similar to that of the circular tombs of Mycenae; and the objects of gold, bronze and ivory, the glass paste, the pottery and the arms—above all, the two splendid golden cups found at Vaphio adorned with representations of combats between men and bulls, afford the most striking analogies with those of Mycenae.* It is the same with recent excavations at Tiryns (1885), where a palace like that of Mycenae has been brought to light on the summit of the acropolis. But this is not all. The remains discovered in the Archipelago beyond the limits of continental Greece belong to the same civilisation. The engraved stones of the islands reproduce the ordinary motives of the architecture and the painting of Mycenae; and in the discoveries of Santorin (of objects anterior in date to the volcanic eruption which overwhelmed the island), in the excavations carried on at Rhodes, particularly in the necropolis of Ialysus, as well as in the necropolis of Cnossus in Crete and in that of Arsinoë in Cyprus, and in the ruins of Hissarlik, a multitude of objects have come to light, a comparison with which is singularly instructive. Thus a new group of antiquities has been formed and is increasing every day, which is represented by these six names— Hissarlik, Santorin, Ialysus, Mycenae, Tiryns and Spata;

* "Les Vases d'Or de Vafio" (*Bull.*, 1891).

and a whole epoch in the civilisation of the ancient world has been revealed to us. At all these different points, so far separated one from the other, we discover a state of art and industry hitherto unknown to us, which was common to all the eastern basin of the Mediterranean. Doubtless, this civilisation was not developed in a day; its types, as the sites of the different excavations present them, succeeded each other during a period of several centuries. "Hissarlik marks its remote origin and early development, Santorin its more advanced condition; Cnossus and Ialysus display the perfection of vegetable ornamentation; Mycenae and Tiryns the abuse of these principles of decoration, and the marvellous progress of this art; Spata marks the point at which the Asiatic influence which had long been felt, became preponderant."* Two streams of tendency, in fact, run through all these civilisations: on the one hand there is an indigenous and local art whose forms do not change, whose decorative principles, whilst perfecting themselves by long practice, nevertheless retain their specific character; on the other hand a foreign influence, increasing every day, mingles with these indigenous elements the productions of a more highly developed civilisation and art. Mycenae preserves in its vases the shapes of Santorin and Ialysus, but there is greater variety and more studied elegance in their ornamentation, and Oriental influence is more marked. But let us compare Mycenae with Spata. If certain shapes are still akin, if certain types are analogous, the general aspect is entirely different. The relative dates too of these civilisations are incontestable. Hissarlik is

* A. Dumont, *Les Céramiques de la Grèce propre*, i. 69.

older than Santorin and Ialysus, while these precede Mycenae, which in its turn is older than Spata. If any one insists on fixing approximate dates, we may place Hissarlik before the sixteenth century, Santorin in the sixteenth, Ialysus in the fourteenth, Mycenae and Tiryns in the thirteenth or the twelfth, Spata in the eleventh century, before the Christian era. No doubt it is difficult to establish well-defined relations between this Mycenean civilisation and strictly Hellenic art. Mycenae marks almost the last stage of a long artistic development; and between the period of its greatness and that in which Greece became conscious of her own powers, a formidable convulsion, the Dorian invasion, swept over the soil of Greece. What matters it? In that remote civilisation there was an independent effort which should form a natural introduction to the history of Greek art. The discoveries of Mycenae were at first sight like a strange and unknown land; to-day they form the earliest chapter in the history of Greek antiquities.

CHAPTER II.

THE EXCAVATIONS OF TIRYNS (1884-5).

Books of Reference:
        The same as Mycenae, together with
  Schliemann, *Tiryns.*
  Marx, "Der Stier von Tiryns" (*Jahrbuch d. d. Arch. Inst.*, Vol. IV.).
  Perrot, *Tiryns* (*Journal des Savants*, 1890).
  Prof. Middleton, *Jour. Hell. Studies*, Vol. VII.
  Prof. Gardner, "The Palace of Tiryns" (*New Chapters in Greek History*).

At the southern extremity of the plain of Argos at a short distance from the sea, in the midst of a strip of marshy lowland, there rises a rocky height which commands the road from Nauplia to Argos. On this plateau, scarcely eighty-five feet high, there rose of old the ancient and powerful citadel of Tiryns, whose mighty walls passed for one of the wonders of ancient times, and seemed worthy, by their colossal mass, of being compared with the pyramids of Egypt. Legend attributed these gigantic structures to fabulous architects, the Cyclops from Lycia, and told how mighty kings had formerly reigned over this fortress; according to the myth one of the most famous of Grecian heroes, Hercules, was born within its walls and long bore rule there. Nevertheless the great name of Mycenae had early eclipsed its humbler neighbour, and Tiryns accepted the lordship of the princes who reigned in the lofty citadel of Agamemnon, and seems to have shared the fate of its masters until the day on which the jealousy of Argos overthrew both suzerain and vassal at one blow. From that time Tiryns was deserted, and gradually

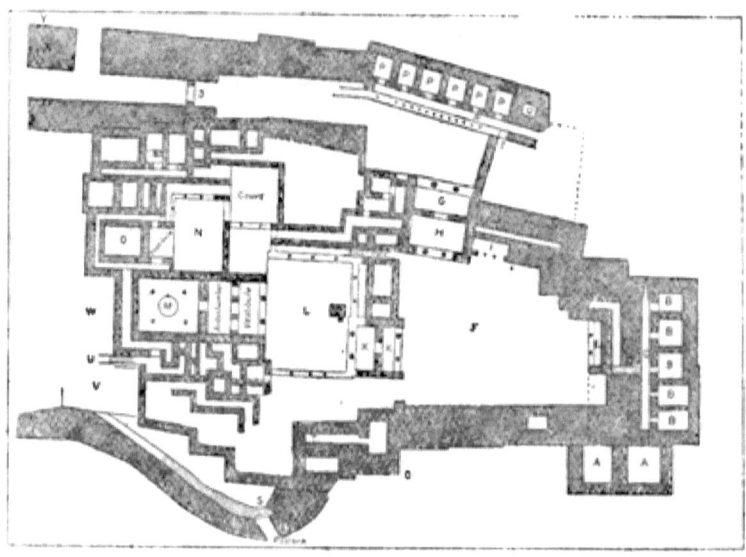

Plan of the Upper Citadel of Tiryns.

A Tower in two Chambers.
B Vaulted Chamber.
C Gallery.
D Corridor and Staircase.
E Portico.
F Large Forecourt.
G S.W. Corner of the Palace.
H Great Propylaea.
I Portico.
K Little Propylaea.
L Large Courtyard.
M Men's Megaron.
N Small Courtyard.
O Megaron of the Women.
P Vaulted Chambers.
Q Cistern.
R Gallery in the E. Wall.
S Side Entrance.
T Cellar-like Rooms.
U Small Staircase.
V Entrance to Middle Citadel.
W Middle Court.
X N.E. Tower.
Y Ramp of Main Ascent.
Z Gate of Citadel.
1 Altar.
2 Door to Gallery.

buried under heaps of rubbish; nothing was to be seen on the summit of the plateau, among the remains of a few Byzantine buildings, but a confused heap of calcined walls which seemed to date from the Middle Ages. The enclosure walls, overthrown, or partially buried

THE EXCAVATIONS OF TIRYNS. 43

beneath the soil, had lost their former imposing aspect, and the travellers coming from the powerful citadel of

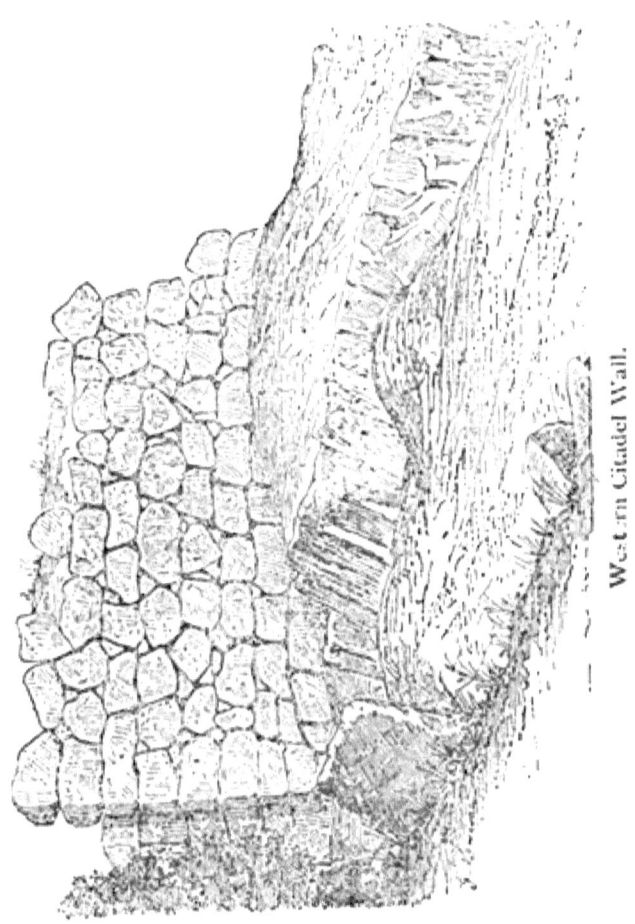

Mycenae, which rises so proudly on its rocky base, had scarcely a glance for the ruins of Tiryns, where scarcely anything seemed to attract their attention. Nevertheless

the enormous size of the blocks used in the buildings, the remains of subterranean galleries curiously constructed in the thickness of the walls, and a few isolated columns of an unknown building on the summit of the acropolis, had already drawn the attention of archæologists to the old fortress, when the excavations recently undertaken by Schliemann came to bestow on it fresh celebrity.

Trial shafts were sunk in 1876, but the work was almost immediately abandoned, and it was not until 1884 that Dr. Schliemann resumed the excavations; and here again, during two successive campaigns (1884, 1885), fortune crowned his efforts with surprising success. Not only has the ancient wall of the citadel, carefully freed from the *débris* which covered it, been restored to us, with its towers, its posterns, and its subterranean galleries, thus revealing to us the skilful and complicated system of early fortifications, but on the summit of the acropolis an extensive palace has been discovered amidst the ruins, with its propylaea ornamented with columns, its courts surrounded by porticoes, its vast halls decorated with paintings; and the plan of this building, restored with the utmost exactitude down to its smallest details, gives us a living picture of the splendid palaces of Alcinous and of Menelaus, described in the Homeric poems. Dr. Schliemann, who liked triumphant telegrams, announced his discovery to the world in the following terms: 'Three cheers for Pallas Athene! I have indeed worked here with surprising success. A vast palace with innumerable columns has been brought to light. The wall-paintings are of the greatest interest, and the vase-paintings not less so—the most primitive representations of plants and animals. The plan of this wonderful prehistoric palace can be exactly restored, and its discovery, which has no parallel, will arouse general admiration."

At first, we must confess, it aroused some scepticism. Eminent archæologists, among them the distinguished English architect Penrose, maintained that the so-called Homeric palace was only a Byzantine building of the tenth or eleventh centuries. Schliemann's discovery was treated as an "extraordinary hallucination of an unscientific enthusiast"; and the *Times*, which took part in this warfare, made merry over this result of an archæological crusade the triumph of which had been announced to the world with a great flourish of trumpets. Fortunately Dr. Schliemann, warned by his experiences at Mycenae, and wisely mistrusting his adventurous imagination, had taken all possible precaution at Tiryns. He had secured the assistance of a man of proved scientific merit, Dr. Wilhelm Dörpfeld, now the director of the German Archæological School at Athens; and this distinguished architect had conducted these excavations on a perfectly trustworthy and scientific plan. Entrusted in 1885 with the completion of the excavations, he had, in the work published by Schliemann, given an account as exact as it was remarkable of the discoveries which had been made; and his testimony is worthy of all confidence. This was plainly seen when, in 1886, the Hellenic Society held a solemn debate in London on the question: Doctors Schliemann and Dörpfeld, both of whom came from Athens for the purpose, had no difficulty in convincing their opponents.

About the same time also, the remains of a palace similar to that of Tiryns were discovered on the summit of the acropolis of Mycenae. It is constructed according to the same plan and after the same methods, is decorated with paintings of the same style, and dates from the same time. To-day all opposition is at an end, and we may fearlessly admire at Tiryns a fortress and a palace erected some thirteen or fourteen centuries before our era.

I.

The citadel of Tiryns rises on a rocky platform 328 yards long and 109 yards broad, which slopes slightly from south to north, so as to form three successive terraces. On the highest terrace, from which there is an uninterrupted view over the sea towards the south, was built, in the finest situation in the fortress, the palace of the king; a little lower, in the middle of the fortress, another terrace about 33 yards broad, and connected with the palace by a narrow staircase, served to lodge the retinue and a part of the garrison, below this stretched the lower citadel, long and narrow, which doubtless contained the storerooms and the stalls for the horses; here, however, the excavations have not yet been completed. The plateau is entirely surrounded by thick and lofty walls built of great blocks of roughly hewn limestone, some of which are of colossal size, measuring between six and ten feet in length, and more than three feet in height and thickness. Although very roughly put together, and much more archaic in appearance than the polygonal stones of the circuit-wall of Mycenae, these blocks were not, as was long believed kept in place by their weight alone; in the interstices, filled up by smaller stones, traces have been found of the

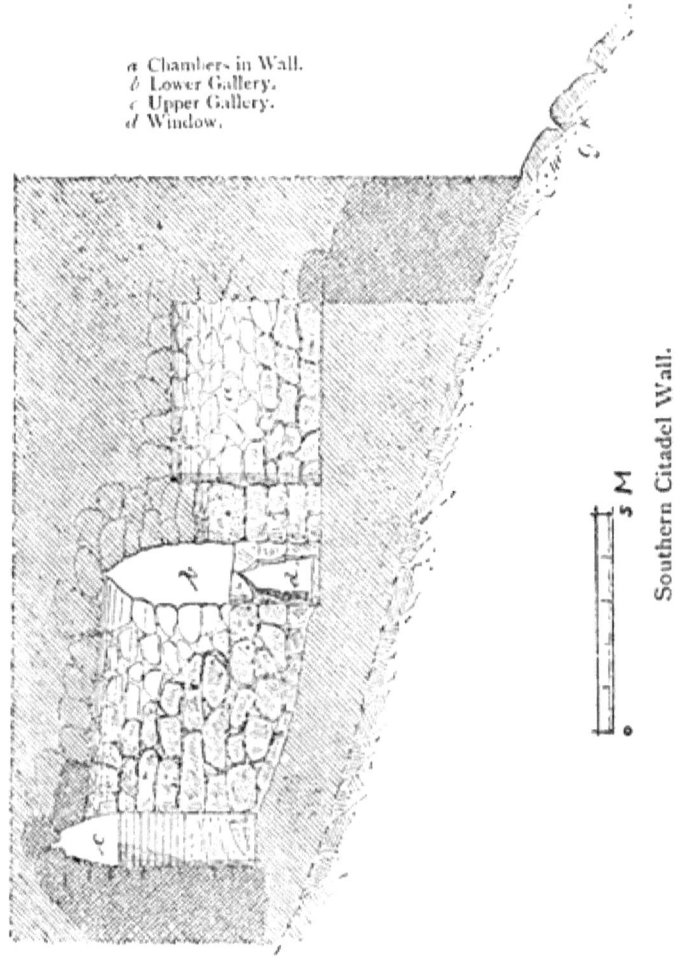

a Chambers in Wall.
b Lower Gallery.
c Upper Gallery.
d Window.

use of mortar. In this way a gigantic wall was built round the whole circumference of the plateau, 23 to 26 feet in thickness, and at least 29 feet high in the lower citadel ; in the upper citadel its proportions were still more formidable. In the latter the walls reached a thickness of 49 to 57 feet, while the height was not less enormous ; the most powerful means of defence were here accumulated round the palace, which formed the centre of the citadel ; numerous projecting and re-entering angles broke the outline of the fortress ; enormous towers covering the principal entrances were thrown across the wide roadway ; a large semicircular projection formed a kind of exercising ground on the western side, and on the top of the wall, a colonnade, forming an upper covered passage, sheltered the defenders. But undoubtedly the most remarkable part of the fortification consisted of the subterranean galleries, which on the south and east were introduced into the thickness of the wall. It has long been known that the southern wall contained two galleries parallel to the line of defence, built on different levels and probably communicating with one another ; a similar gallery also existed along the eastern wall ; and all three, roofed with mighty overlapping blocks of limestone, seemed intended to give access to the lower parapet of the citadel. Recent excavations have shown the inadequacy of this explanation by laying open a much more complicated network of galleries than had been known to exist. As had been supposed, a staircase connected the upper with the lower gallery ; but the latter, it was discovered with astonishment, opened into five chambers with pointed vaulted ceilings constructed in the thickness of the wall ; in the same way the gallery of the eastern wall gave access to six chambers of this kind, while a narrow loophole in the wall gave the necessary light to the

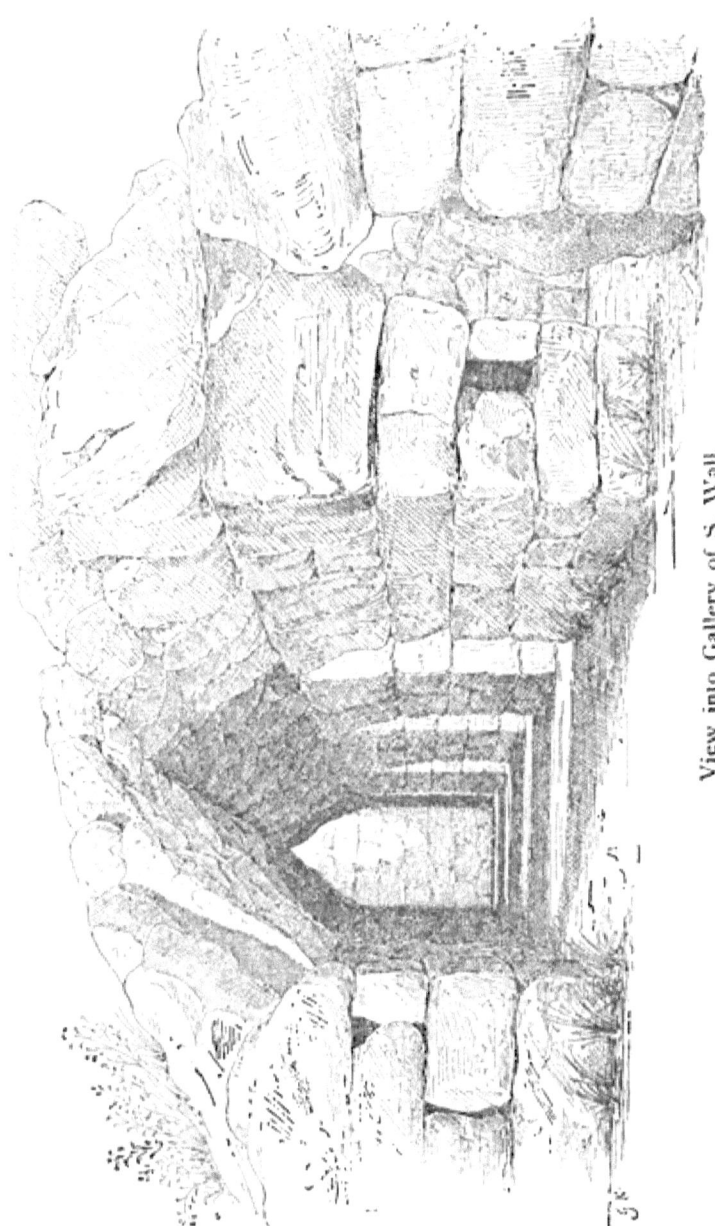

View into Gallery of S. Wall.

corridor. Other subterranean rooms were constructed in the interior of the great towers of the western wall, and thus a whole series of casemates were arranged, which served as cisterns and subterranean storehouses. Lastly, special fortifications protected the royal palace on the north and east, in such a way as to separate it from the lower citadel, and to afford a powerful defence to the principal entrance of the fortress.

Two gates, as was usual, gave access to the inclosure; on the western side a postern in the semicircular projection led to a narrow passage practicable for foot passengers only, which led to the middle citadel by a staircase, sixty-five steps of which are still visible, and thence ascended by a few steps to the palace. This was a secret exit on the side towards the sea, and played a very important part in the defence of the place. The principal approach, which served for wheeled vehicles and horses, lay on the eastern side. In order to mount gradually to the level of the fortress, 65 feet above the plain, the road ascended by a long ramp to the foot of the circuit-wall, whose towers and projecting angles commanded and covered it. It debouched at the summit in a narrow passage opened in the wall, and flanked by a colossal tower; it then ran between the eastern wall of the citadel and the fortified terraces of the palace, and led to the gate of the fortress, which was very similar in its size and construction to the lion gate of Mycenae. We can still see the heavy stone threshold and the enormous posts which formed this doorway; and traces have been found of the heavy beam which, slipped through the uprights, served to bolt the door. Even this was not all; for, this obstacle surmounted, a narrow passage, sixty feet or more in length, still remained to be traversed ; then at last towards the south-east angle of the fortifications, the road terminated in a wide open space, on the opposite side of which was an imposing doorway. This was the entrance of the royal palace.

We see with what extreme care everything had been foreseen which was necessary for the defence ; with what prudence every access to the fortress had been protected, and the enemy compelled to advance for a long distance under fire, through narrow passages commanded by high walls and covered by strong towers ; and with what skill the projecting angles of the walls had been constructed and the terraces raised one above the other, so as to make the palace a kind of donjon within the fortress itself. The methods of construction, too, however rude they may be, bear witness to a workmanship which is already skilful and master of the means at its disposal. Experienced architects were needed to bring into position the blocks of the wall—the smallest of which weigh from three to four tons, and some as much as thirteen ; as well as workmen, already masters of their trade, to vault these galleries and casemates ; above all, it needed an army of labourers, long years of peace, great resources and enormous riches, to construct this formidable stronghold.

The purely defensive works ended at the first gateway and there the palace began. In the latter the buildings had to meet new conditions : vast courtyards and spacious halls were needed there, easy communication between the different apartments, light and especially abundance of shade and coolness, the elegance in fact and luxury which would make the palace worthy of the king who dwelt in it. Until now we had little more than the material afforded by the Homeric poems and the descriptions given there of the palaces of Ulysses and of Priam, of Alcinous and of Menelaus from which to picture to ourselves the palaces of the heroic age, and in spite of ingenious efforts to reconstruct the plan of these buildings, many questions remained unanswered. Now, upon the acropolis of Mycenae or of Tiryns we can admire the design of the old Homeric palace in its full development. These princely dwellings, unearthed by the pickaxe of the excavator, are unquestionably older and in some respects more luxurious than those described in the poems; but the main features of the design, the general principles of the building, are the same, and it is this which gives so much interest and importance to the recent excavations at Tiryns and Mycenae.

Now let us pass through the great propylaea opening on the "terrace," the ground plan of which remained unchanged from heroic times down to the erection of the Propylaea on the Acropolis at Athens. We enter a large, almost rectangular courtyard, surrounded by chambers and porticoes, from which the stairs lead down to the subterranean galleries in the wall; crossing this we enter, through a second imposing gateway in the northwest corner called the little propylaea, the palace properly so called. Here we find another large rectangular courtyard surrounded by porticoes, 51 ft. 7 in. deep by 66 ft. 4 in. broad: it is the court of the men's apartments, the highest point and the centre of the whole palace. On the southern side rises the traditional altar of Zeus Herceius, a massive square of masonry below which the circular pit destined to receive the blood of the sacrifices was hollowed out; while facing the altar, on the northern side of the courtyard, is the men's megaron. Ascending two steps, we reach a vestibule adorned with columns, the floor of which is still covered, like that of the court itself, with a kind of plaster; through three large doors we pass into an ante-chamber, and thence into the state apartment, the court of the men or megaron, a large room nearly 40 ft. long by 32 ft. broad, which was the chief apartment of the palace; four columns in its midst supported the roof, and between them upon the ground, which the red and blue patterns of the concrete floor seemed to cover with a kind of Oriental carpet, was the hearth.

To the left of this great hall were a number of corridors and small rooms in direct communication with the court or megaron, among which an interesting apartment has been recognised—the bath-room. The importance attached in the early ages of Greece to cleanliness, and the attentions of this kind always offered to guests are well known, and account for the situation of the bath-room, near the ante-chamber and not far from the reception-room. The arrangement of the room itself is very remarkable: the floor is made of a huge block of limestone weighing no less than 20 tons; the walls were formerly panelled with wood; in the middle was a bathing-tub of terra-cotta and the water ran off through underground pipes constructed below the floor of the palace in such a way as to collect into one large drain the surface water of the courtyards and of the house.

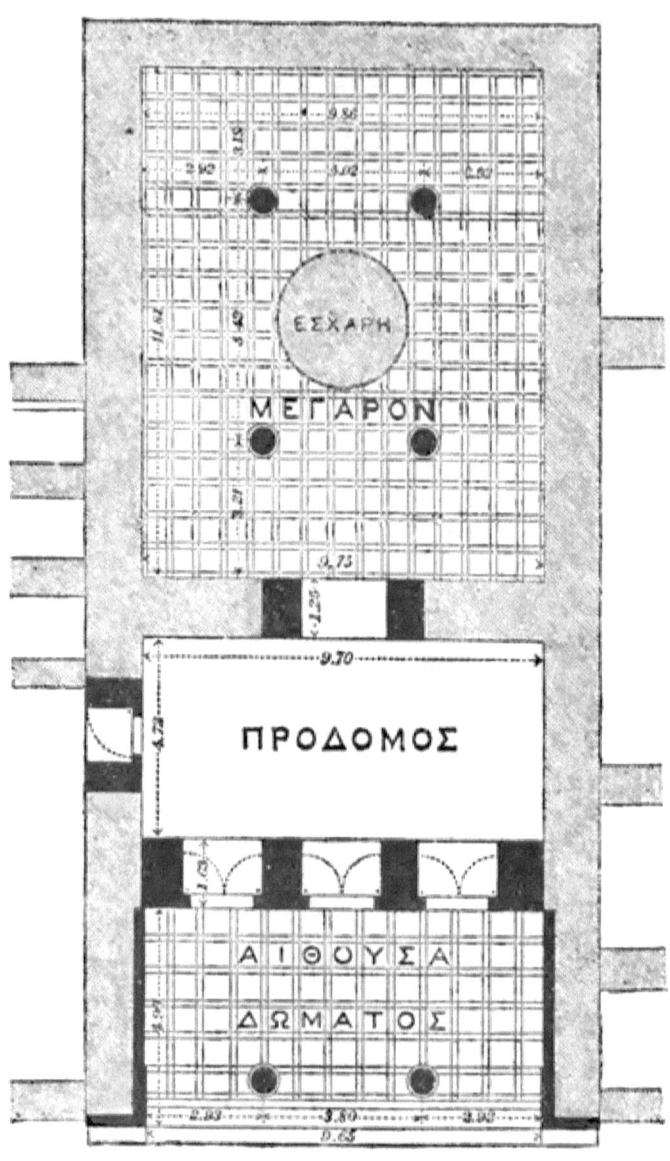

Plan of Men's Apartment.

To the right of the men's was the women's apartment, which was also entered through a courtyard surrounded by porticoes, and which consisted of a vestibule and a rectangular hall like that of the megaron, but smaller. The women's apartment, which had no direct communication with the rooms set apart for the men, was cut off from the rest of the palace by a succession of doorways and passages. It was connected on the one hand with the great propylaea by an inner courtyard and a long corridor, on the other with the great courtyard of the megaron by a series of passages which went completely round the men's apartments and ran through the whole left side of the palace. In this, the most retired and the most isolated part of the palace, was situated the sleeping apartment of the royal couple, entered through an antechamber, as well as the arsenal and the treasury. Such is the plan of the palace of Tiryns, but we must set up the fallen walls again on their stone foundations, which are almost all that remains of them; we must rebuild the roofs that have fallen in: we must recover the decoration of these halls and the main outlines of their arrangement, in order to restore to this royal dwelling its former aspect and animation. For the most part the walls of the palace of Tiryns were built of unhewn blocks of limestone bonded with clay-mortar, and long wooden beams were fitted into the masonry across its breadth, which played the part of chains and gave more solidity to the building. Unburnt brick was often used in the upper parts of the walls, and the whole was covered with a coating of clay and plastered with lime. Wood in especial played an indispensable part in the building.

The corner-pieces which sustained the facing of the walls, the door-posts, the pillars and their capitals, the pilasters, the ceiling, sometimes even the thresholds of the rooms, were of wood ; and wooden panels covered the walls of certain apartments, such as the bath-room and the vestibule of the megaron. The roofs supported on squared beams were horizontal, and arranged in terraces after the fashion of the East. Light was admitted through the doorway, and by means of lateral openings in the upper part of the wall under the roof.* Lastly, the halls were carefully paved with a kind of mosaic, in some places resembling a carpet, and decorated with mural paintings and carved ornaments, of which some very interesting fragments have been recovered. One of the most remarkable is a charming alabaster frieze, inlaid with blue glass-paste, which ornamented the vestibule of the megaron, and reproduced patterns already familiar at Mycenae, at Menidi, and at Orchomenus. The frescoes on the walls are not less interesting, and throw a curious light on the early history of painting in Greece. Sometimes wide parallel stripes of different colours are painted upon the coating of lime ; sometimes we find flowers, rosettes, meanders, spirals, the arrangement of which offer striking analogies with the sculptured ceiling found in the beehive tomb at Orchomenus. In other places the human figure begins to appear ; a quantity of fragments have been discovered, forming part of those winged monsters so dear to the imagination of early artists, but the most remarkable of these paintings is that which represents a furious bull, upon whose back is a man, half-kneeling.

* *Tiryns*, p. 129. See Professor Middleton, J.H.S., vol. vii., p. 161.

"A Furious Bull, upon whose back is a Man, half-kneeling."

It must be admitted that the technique of these frescoes is still very rude. The painter's palette has only five colours—black and white, blue, red and yellow—but their effect was none the less striking when they shone with all the brilliancy of their simple colouring.

Nevertheless, we must not allow ourselves to be dazzled by the seeming splendour of this decoration. The Homeric palaces, of which that of Tiryns is the model, were destitute of many refinements of comfort and cleanliness, which seem indispensable to us to-day. Cooking was carried on all day on the hearths which form the centre of the principal apartments, and the smoke covered the walls with a thick coating of soot; the smell of burning fat floated in the heavy atmosphere, and the offal of the beasts, as they were slain, the heads and feet of oxen, and hides wet with blood, were dragged aside into the corners.

Before the very door of the house there was a heap of dung, on which lay the dogs, infested with vermin. It is possible that the civilisation of Tiryns, though more ancient, was more refined than this—for the Homeric palaces possessed neither the wall paintings nor the paved floors which we find on the acropolis of Tiryns, but the general effect must have been much the same. In order to bring before ourselves a true picture of what these palaces were like, we need only think of the condition, in modern times, of the 'konaks' of the pachas and beys of Asia Minor. "Wood and stone are mingled there in the same way throughout the vast extent of the buildings; the 'selamlik' corresponds to the megaron, the harem to the women's apartment; in the vast and filthy courtyards, surrounded by the lodgings of the slaves and the storehouses for provisions, there is the same medley of idlers loitering and animals wandering freely about; in the interior of the dwelling there is the same contrast of luxury and neglect. Thus the life and habits of the East in modern times still furnish in many respects the best commentary on the Homeric age."*

II.

The discoveries made at Tiryns—and this is not the least part of their interest, complete in a very curious way the picture of that early civilisation which was revealed by the excavations of Mycenae. The buildings and remains found on the one and the other acropolis date, in fact, from the same time, and belong to the same period of greatness and glory. We have already pointed out the striking analogies which exist between the paintings of Tiryns and the carved ceiling of Orchomenus. There is no less resemblance between these frescoes and those which decorate the palace of Mycenae, where we again find the long parallel bands of different colours, and the timid attempts in which the artist endeavours to represent the shapes of men and animals. The pottery found at Tiryns is on the whole in the same style as the vases of Mycenae; the frieze of the palace is exactly similar in design to the patterns in use both there and at Menidi; lastly, Oriental influence has left its traces on both,—and this is not the least interesting of the lessons taught by the discoveries at Tiryns.

The traces of Phoenician influence are visible every-

* Perrot, *Revue des Deux Mondes*, July 15th, 1885.

where, both in the methods of construction and in the style of the decorations. The subterranean galleries contrived in the thickness of the walls, together with the vaulted chambers, beside them, are met with again in the fortifications of the Phoenician cities of Africa, at Carthage and Thapsus, at Hadrumetum and Utica, and the resemblance between the two systems of fortification is so great that we must consider it an undoubted fact that Phoenician architects, or at all events architects trained in Phoenician methods, were present at Tiryns. It is from Phoenicia, too, that the technical processes employed in the construction of the palace walls were derived; the Phoenician workmen who built the temple at Jerusalem for Solomon interspersed the courses of masonry with beams of cedar, and the same mingling of wood and stone is found both at Tiryns and Mycenae. Lastly, the blue glass-paste, with which the alabaster frieze of the megaron is inlaid, is an Oriental importation. This method of decoration was in use in Egypt from the most remote antiquity, but she received the raw material from Phoenicia, and it was from Phoenician manufactories that those glass-pastes, coloured with copper-salts, were brought into Greece, which are found at Menidi and at Spata, at Mycenae and at Tiryns, with which Homer, too, was acquainted, and which he speaks of under the name of kyanos as decorating the frieze in the palace of Alcinous.*

It is impossible then to overlook the Oriental influences which have so deeply affected the civilisation of Tiryns. The palace of Tiryns, too, is of later date than the graves

* *Od.* vii., 87.

of Mycenae. A single day did not give birth to the long series of monuments which mark the growth of this remote civilisation; the graves cut in the rock of the acropolis at Mycenae are older than the beehive tombs in the lower town and the palace on the summit of the hill. The passage-tombs of Spata and Nauplia are still later than these latter remains. The general features are everywhere the same, but the details differ infinitely as the centuries pass on; art becomes more certain of itself, and Oriental influence grows more powerful, until the day arrives on which a new civilisation, represented by the Athenian Dipylon vases, replaces the older culture of Mycenae. The palace of Tiryns is contemporaneous with the beehive tombs of Mycenae, but it seems to have lasted longer than the Mycenaean buildings. Some fragments of pottery in the geometrical style, ornamented with human figures exactly resembling those of the Dipylon vases, have been found amongst its ruins, as well as some still more curious fragments which seem to form the transition between the Mycenean vases and those of the next period. Thus Tiryns shows us, after the greatness of the Mycenaean civilisation, its closing efforts and its decadence.

When the Dorian invasion passed over Argolis all that remained of this brilliant civilisation was swept away by the storm. A ruder race replaced the wealthy and luxurious princes who reigned at Tiryns and Mycenae; and long years of warfare rendered their manners still coarser and more barbarous. At the same time the Phoenicians who had been the teachers of primitive Greece, were gradually forgetting the way thither. Banished from the islands of the Archipelago one after another, deprived of the monopoly of sea-borne trade, driven away by the incessant warfare which destroyed all security, they no longer maintained the same close relations as formerly with the western shores of the Aegean. Civilisation could not but suffer by the change, and in fact the Homeric is far more barbarous and uncivilised than the Mycenaean period. Still, even then the gains of the past had not been wholly lost, and the palace of Tiryns is still the model in wood and stone of those of the heroic age. It is on account of this connection that the building so closely concerns the history of human dwellings, for the Greek house is merely a development of the Homeric palace such as we see it in the excavations on the acropolis of Tiryns.

CHAPTER III.

THE EXCAVATIONS AT DODONA (1876).

BOOKS OF REFERENCE:
  Carapanos, *Dodone et ses Ruines*. Paris, 1878.
  Girard, "Dodone et ses Ruines" (*Revue des Deux Mondes*, Feb. 1879).
  Bouché Leclercq, *Histoire de la Divination dans l'Antiquité*.
  Hoffmann, *Orakel-inschriften aus Dodona*.
  Roberts, "Oracle Inscriptions discovered at Dodona" (*Journ. Hell. Studies*, vols. i. and ii.).

THE scenery of the natural highway which passes over Mount Pindus from the Greek province of Thessaly into the Turkish province of Epirus, and which has from ancient times been the means of communication between the Adriatic and the Aegean, offers a remarkable contrast to the traveller. On leaving Larissa he ascends the lofty valley of the Peneus through forests of oak and plane-trees, and passes at the foot of the strange monasteries of the Meteoroi, clinging boldly to the summits of the lofty peaks of basalt, and hanging, as it were, between earth and heaven. As the road winds up the mountain side, his eye sweeps over the whole extent of the fertile plain of Thessaly; but no sooner has he passed through the defile of Mezzovo than the whole scene is changed. Around him are bare and rugged slopes, a chaos of mountains rising confusedly one behind the other; in the words

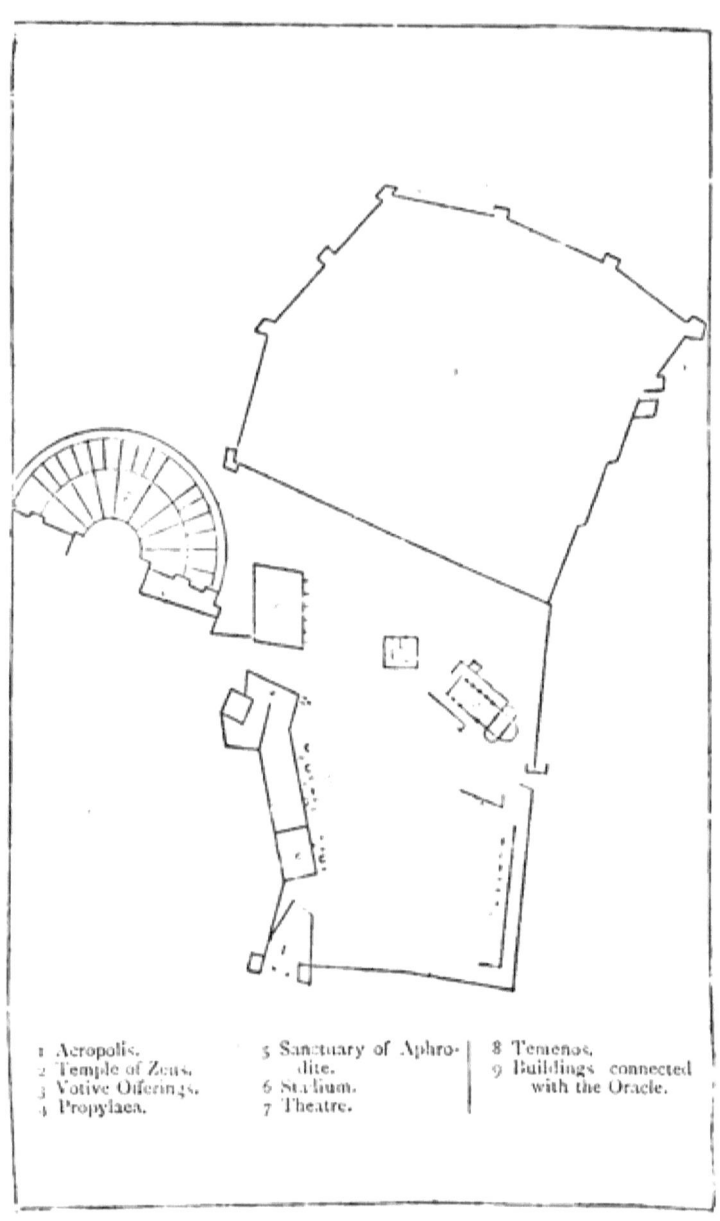

1 Acropolis.
2 Temple of Zeus.
3 Votive Offerings.
4 Propylaea.
5 Sanctuary of Aphrodite.
6 Stadium.
7 Theatre.
8 Temenos.
9 Buildings connected with the Oracle.

Plan of Dodona.

of M. Girard, Nature here is harassed and ungenial, and cannot expand freely in any direction. Even the lake, which lies at the foot of the mountain and reflects in its waters the citadel and minarets of Janina, scarcely softens the severity of the scenery; and elsewhere on all sides the lofty peaks are separated by deep valleys which are covered with marshes, or form the beds of mountain torrents. The climate is as ungenial as the scenery. During a great part of the year the mountains are covered with snow, the wind howls through the narrow passes, and the thunder rolls around the peaks. In some of the valleys of Epirus the average number of stormy days in a year is forty-nine—the highest in all Europe.

It was in the midst of this gloomy scenery that Greek mythology localised some of the legends of the lower world, and it was here too, where Nature exercised a powerful influence upon the fresh unhardened minds of the nations of antiquity, that one of the most ancient and venerated of the shrines of Greece arose. In the cold and narrow valley of Dodona, under the shadow of the great oaks which surrounded the sanctuary of Zeus, Greek mythology received its earliest form, and the first of the great Greek oracles arose. Under this rigorous sky there grew up a stern, harsh worship, some of whose observances seem to savour more of the asceticism of the Middle Ages, than of the joyous freedom of the ancient world. It is this individuality which lends interest to the sanctuary of Dodona, and makes the history of its oracle stand out so clearly against the usual colouring of the ancient world, as to deserve something better than oblivion.

I.

Until the last few years the sanctuary of Dodona was scarcely more than an illustrious name. Even its exact site was unknown, though history and legend were full of its fame. In *Childe Harold* Byron writes:—

> "Oh, where, Dodona, is thine aged grove,
> Prophetic fount and oracle divine?
> What valley echoed the response of Jove,
> What trace remaineth of the thunderer's shrine?"
>
> *Canto* ii. 35.

It was well known that Dodona was in Epirus, and it had been sought in the neighbourhood of Lake Janina; but until the day on which in 1876 M. Constantine Carapanos, in the course of his memorable excavations, actually discovered the sanctuary itself, the most complete uncertainty prevailed as to its site. To-day it is beyond a doubt. To the west of Lake Janina, in the moist and fertile valley of Tcharacovista, at the foot of the barren peak of Tomaros, situated in a picturesque and gloomy spot, M. Carapanos has discovered the ruins of the oracle of Zeus. The narrow circuit-wall of the ancient citadel, studded with towers, rises on an out-lying height commanding the plain, in the midst of a mountain valley between 1600 and 1700 feet high. At the foot of the acropolis a fine theatre, still in a perfect state of preservation, served to celebrate the games in honour of the gods of Dodona; and on the slope of the hill, close by it, the walls are still standing which mark the boundaries of the sacred enclosure, in whose midst, on the summit of the plateau, there rose the temple, afterwards transformed into a Christian Church. M.

Carapanos has succeeded in tracing the ground plan of the sanctuary in its main outlines, as well as that of two adjoining buildings, which no doubt contained the treasure of the god, or were in some way connected with the service of the oracle. Below these buildings there extended a large open space, entered through a vestibule of considerable size, and bordered by porticoes along which the ex-votos offered to Zeus by the piety of the faithful were arranged on rows of little pedestals. In the centre of the square there doubtless stood the grove of sacred oaks to which Dodona owed its fame, and which concealed within its recesses the prophetic tree of Zeus, the miraculous spring, and the oracle of the god. M. Carapanos has made more than one interesting discovery among the ruins of the ancient sanctuary. It is true that he has not found an extensive series of marble statues—marble was rare in Epirus—nor any collection of treasures in gold or silver, for during its latter years the temple was plundered more than once, and even completely destroyed; and the traces of a great fire which laid it waste are still visible among the ruins. Nevertheless,

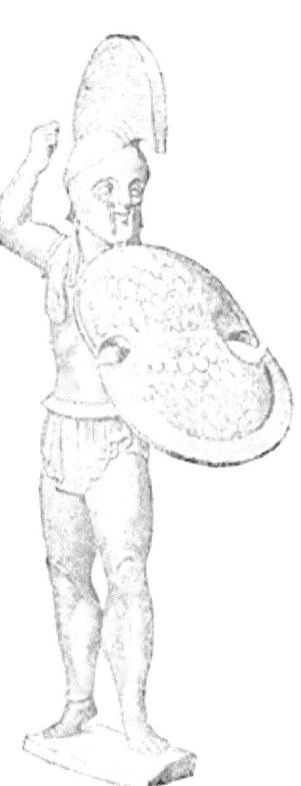

Bronze Statuette from Dodona.

if he has not found any objects of great value, he has discovered a multitude of ex-votos and pious offerings, vases or candelabra of bronze, tripods and wreaths, and a quantity of trifling articles of daily use—arms and ornaments, spurs and spear-heads, bracelets and earrings, hairpins and mirrors, and even those small spatulas used by women for applying paint. Just as at the present day the worshippers at famous shrines hang gold and silver hearts or images of saints on the walls of their chapels, so they offered to the gods in ancient times statuettes of bronze, or of terra-cotta; and many of these bronzes found at Dodona, whose beautiful green or blue patina strikes the eye at once, are of the greatest importance for the history of archaic art. The series of small leaden plates, however, from the temple archives, on which the pilgrims, who came to consult the oracle, engraved the words of the questions they asked, are of still greater interest. Although only a small number of these fragments have come down to us—scarcely forty-five have been found—they undoubtedly form the most interesting part of M. Carapanos' discoveries. Thanks to these unique documents, we can penetrate behind the scenes in these oracular shrines, and are admitted to the privacy of religious life in the ancient world. It was in this remote district of Greece, in "wintry" Dodona, as the poet calls it, in the midst of the vast woods and the murmuring fountains, among all the mysterious sounds of nature, that the Greeks first thought they heard the voice of the gods, and that we must look for the cradle of Greek religion and its earliest manifestations.

II.

The pious pilgrim, when he reached Dodona, found that both the worship and the priestly body there were twofold: with the cult of Jupiter was associated that of a goddess, Dione; and by the side of the priests called Selli or Tomouri there existed a body of priestesses, known as Peleiades or doves. What was the origin of these divinities and of the priestly bodies devoted to their worship? The question already occurred to Herodotus, who visited Dodona at the end of the fifth century, and made careful inquiries of the priests concerning their sacred legends. Such inquiries caused no embarrassment in the case of any ancient cult, and the priestesses of Dodona related a very interesting story to their inquisitive visitor. It always seemed a pleasure to the ancient Greeks to connect their oldest traditions with Egypt; the powerful monarchy of the Pharaohs not only gave the tone in art and fashion, but was the object of their admiration in religion as well. Accordingly it was in Egypt that they sought for the remote origin of the shrine of Zeus. Two doves, they said, flew away one day from the city of Thebes in Egypt, and, like the pigeon in the fable, took their flight into distant lands. One alighted in Libya, on the spot where the oracle of Jupiter Ammon was afterwards established; the other, crossing the sea, flew as far as Dodona, and, perching on an oak, uttered in human voice the command to those who heard her to establish there an oracle of Zeus. Such was the story told to Herodotus by the three priestesses of the temple; but the old historian, in his naïve zeal for the solution of religious problems, wished to learn more. When in the course of his travels he chanced to arrive at Thebes, in Egypt, he consulted the priests again; and this time he was told that two priestesses had been carried off from Egypt by the Phoenicians and sold into slavery, the one in Libya, the other at Dodona. "And on my inquiring how they came to know so exactly what became of the women, they answered that diligent search had been made after them at the time, but it had not been found possible to discover where they were; afterwards, however, they received the information which they had given me."* Herodotus, who was not exacting in the matter of belief, thought this sufficient, and with the ingenuity of a Greek contrived to reconcile the two legends. In the black dove of Dodona he gladly recognised the dark-complexioned Egyptian, whose barbarous and unintelligible language had been compared to the twittering of a bird, "a curious example of the candid and reasoning credulity of the Greeks, and of the childish trifles which could still beguile the powerful intellect which at the same time was the first to undertake the historian's task." †

Nevertheless there was a basis of truth underlying these sacred legends. After the usual fashion of primitive races, the first inhabitants of the valley of Dodona had deified and adored the mysterious forces of nature —Zeus, that is to say, as his surname Naius indicates, the water which penetrates and fertilises the earth, and Dione, the productive forces of the earth itself. But

---

* Herodotus, ii. 54-58. (Rawlinson.)

† Girard, "Dodone et ses Ruines," *Revue des deux Mondes*, 15 Fév. 1879, p. 941.

other elements were afterwards added to this primitive conception: the symbol of the dove, closely connected with the oriental cultus of Aphrodite, is an evidence of the importation of foreign rites, an importation which the Egyptian priests pointed out to Herodotus. The primitive goddess, Dione, was confounded with the Syrian deity, and to the Pelasgic priests of Zeus were added a body of foreign priestesses who had followed in the train of Astarte. Hence this double priesthood, of which, without dwelling longer on these mythological mysteries, some details must be given here.

In the religious history of antiquity, the priesthood of Dodona merits special attention. In a passage of the Homeric poems, Achilles addresses a solemn prayer to the national deity of the barbarous and gloomy country of his birth: "King Zeus, Dodonean, Pelasgian! thou that dwellest afar, ruling over wintry Dodona, and around thee dwell the Selli, thy prophets with unwashen feet and couching on the ground." *

These austere observances, which make the care of the body a hindrance to holiness, would not be surprising among the fakirs of Hindostan; but in the Greek world, in the midst of the free and joyous life of the Homeric poems, this stern apparition contrasts strangely with the general tone of the picture. Without these few lines of Homer, we should never have suspected the existence of asceticism in the early days of Greece, and we should be tempted to say that the Hellenic world had never known it. In the same way, as we read the sparkling story of a Froissart, we see nothing in the Middle Ages but a world of elegance and

* Lang Leaf, Myers, *Iliad*, xvi. 234.

frivolity, where all is noise and gaiety, and wears a joyous air till the very day of death, yet, notwithstanding, at that very time was written the *Imitatio Christi*, so full of melancholy and asceticism.*

By the side of these pagan cenobites the priestesses of the temple were subjected to an equally severe discipline. In the gloom and barbarism of Epirus, under the stern sway of the ascetics of Tomaros, they soon lost their former grace. They too had to adapt themselves to the atmosphere of the place; a rigorous system of discipline replaced the laxity of the Syrian temples, and a high-priest, at the head of the double community, was charged with the supervision of the priestly body, lest any external and mundane interests should deprive the deity of their exclusive devotion.

There was still another reason for the brilliant reputation which the sanctuary of Dodona enjoyed throughout Greece. The mysterious breath of the gods passed through the sacred groves surrounding the temple; miraculous voices were heard amidst the murmurs of the spring, which gushed out at the foot of the mighty oaks; the divine will was manifested in the vibrations of the tripods shaken by the wind; and in all these voices of nature the initiated were enabled to distinguish and to interpret the oracles of the king of heaven. The part played by the oracles in the ancient world and the importance they possessed are well known. To a pious Greek the oracle rendered almost the same services which we now require of our lawyers and doctors. Not only did the cities send official

---

* Cf Havet, *Le Christianisme et ses Origines*, vol. i., pp. 16-17.

deputations to the god to learn from him the principles of sound policy and the rules of good government, but the most insignificant private person, if at a loss to know what line of conduct to pursue, submitted his doubts and scruples to the oracle. Some sought the means of preserving their own health and that of their relatives, others asked for advice concerning the rearing of their flocks; merchants inquired whether their speculations would succeed; and an heir asks whether, in the division of the property, he would do well to take the town or the country house. All come to tell the god their private affairs, and often their communications are of a most confidential character. M. Carapanos has found among the ruins of Dodona a whole series of questions addressed to the oracle of Zeus, some of which display an amusing *naïveté* and indiscretion. On one of the small leaden plates, on which the worshipper wrote his question, one worthy man inquires of Zeus whether he has lost the coverlets and pillows, which had disappeared, or whether they have been stolen. Another would like to know, beforehand, whether heaven has other children in store for him besides the daughter he has already; while other inquirers were in still greater difficulties, and their questions must sometimes have been extremely embarrassing to the god.*

Occasionally, indeed, the oracle was in difficulty.

---

* Cf. Carapanos, *Dodone et ses Ruines*, p. 75. The inquirer often did not tell his name; sometimes, it seems, he asked a mental question. One of the few answers preserved is, no doubt, a response to a question of this kind: "This is the oracle I pronounce—you would be mistaken."

The questions the faithful wished him to answer were sometimes real riddles. On one of the leaden plates discovered by M. Carapanos, the worshipper inquires whether he will succeed in the undertaking he contemplates if he acts in the way he proposes. That is all: it was for Zeus to penetrate the intentions of this somewhat incommunicative adorer; it is not for nothing that one is endowed with second sight. In cases like this, the priests, whose duty it was to interpret the divine revelation, needed rare dexterity and remarkable diplomatic skill, to extricate themselves with credit from such inquiries. Unfortunately very few of the answers of the oracle have been preserved among the inscriptions of Dodona. The reason is evident; the worshippers were careful to carry away with them the advice given by the god; but nevertheless, for our sakes, the loss of these documents is greatly to be regretted. We can well imagine, however, from all we know, that the priests were not lacking in ingenuity. They had at their service the dark and obscure style proper to an oracle, in which the goodwill of the faithful could easily find all it wished. Moreover, like wise men, the priests took care ingeniously to regulate the methods of consultation so as not to compromise by awkward answers the credit of the Dodonean Zeus, so lucrative a source of profit to his priests. There were several methods at Dodona of ascertaining the will of the gods. Sometimes it was by listening with religious awe to the rustling of the prophetic oak, sometimes by watching the flight of the doves which, perching on the sacred tree, awoke its voiceless accents. Now they interpreted the murmurs of the miraculous spring, which possessed the singular property of extinguishing lighted torches, and lighting those which had gone out, when they were dipped into its waters. Again, they made inquiry of the bronze basins hanging in the sacred grove around the oak of Zeus, whose echoes, indefinitely prolonged as the sound passed from one to another, gave the answer of the oracle to the questions asked of it. In later times, in order to retain by some ingenious invention the adhesion of the faithful, they invented a new system of divination by bronze. By the side of a bronze basin, upon a column, they placed a statue of a child with a whip in his hand; and whenever the wind rose, the whip, which was made of three small chains of metal, struck the bronze and made it resound. The echoes cannot have died away for some time; for it was common in ancient times to compare great talkers to the basin of Dodona.

Those who had come to consult the oracle did not, however, enter into direct communication with the god. They would have found it difficult indeed to interpret, in a satisfactory way, all the mysterious voices of nature through which the divine will was made known; and on the other hand, there were many reasons why the priests should wish to exercise a close supervision over the rites of divination. Accordingly they gave small leaden plates to the worshippers who were desirous of consulting the god, upon which each wrote his question and generally his name and country as well; but they were on their guard lest any one should put a question which would be too indiscreet—that is, too difficult to answer. There were certain sacred formulæ, certain regular models, suggested or dictated by the priest, according to which the inquirer framed his question. The people of Corcyra, for instance, wished to know how they might re-establish a settled government in their city, which was distracted by civil discord; but the oracle was too wise to accept the question in such a shape. It had no desire to take the part of any of the factions which were then dividing the state, and the ingenious priest discovered a device by which Zeus need not be entangled in the thankless business of politics. He dictated to the Corcyrean envoy the following question, which is not so difficult to answer. " To which of the gods must prayers and sacrifices be offered, that peace may be restored ? " The gain was twofold, for Zeus was no longer in danger of falling out with the republic, and, what was more, he could, like a courteous deity, direct the faithful to the altars of another god, thus carrying on a very profitable exchange of civilities with the neighbouring shrines, which, without compromising his credit, might bring him a windfall in his turn. The leaden plate was next committed to the priestesses of the temple, whose special office it was to interpret the meaning of the divine manifestations. Formerly the priestesses themselves communicated the answer of the god to the inquirers, but in later times it was found unseemly to bring the women engaged in the service of the temple into direct contact with worshippers whose temper was somewhat rough and coarse ; and from that time the priests became the interpreters to the people of the will of Zeus, and dictated to them the oracles given by the priestesses.*

In order to show their gratitude to Zeus for his kind-

* Bouché Leclercq, *Histoire de la Divination* p. 297.

ness and his good advice, the faithful before carrying home the answer of the oracle, vied with one another in zeal and liberality. Numbers of pious offerings were placed by them in the treasury of the temple, or solemnly sent as thank-offerings to the great god of Dodona. A certain number of these ex-votos have been found in the course of the excavations: tripods of bronze, vases of the same metal, jewels and articles of the toilette, all gifts bestowed by the obscure worshippers who flocked to the shrine. The offerings of the cities were more magnificent; Athens, which was always specially devoted to Dodona and even in the last days of its existence as a free state zealously consulted the oracle, sent a solemn embassy with great pomp, to bear a complete set of ornaments for the statue of the goddess Dione. Princes like Alexander bequeathed considerable sums to the sanctuary. But by the side of these contented clients some, it might be feared, would be found who were dissatisfied. During his prolonged career as a prophet Zeus sometimes made mistakes, and those who suffered from his ill-advice had occasionally the bad taste to complain. Consequently, it was found desirable to preserve carefully in the sacred archives the text of the questions put by the faithful, that it might be produced in case of necessity, so as to prove by documentary evidence that they had not expressed themselves clearly and had only themselves to blame for their misfortunes. It is to this custom that we owe the leaden plates discovered by M. Carapanos. Nevertheless, a choice was made among this mass of documents : the more important were put on one side ; and on busy days the old plates of no particular interest were given out again to the faithful. This is the reason why some of the plates found at Dodona bear several inscriptions.

III.

The discoveries of M. Carapanos are sufficient to prove the high reputation of the oracle and the variety of people who consulted it. In the inscriptions discovered at the shrine we meet with the names of great states like Corcyra and Tarentum, and of private individuals from all parts of Greece, Dorians and Aeolians, inhabitants of the coasts of Epirus and of the neighbouring islands, Athenians and Spartans, men and women of every tribe and every rank, shepherds and merchants, sailors and labourers. Still, an oracle which valued its reputation could not for its own credit's sake be satisfied with such obscure patronage; and just as in some hotels the names of illustrious visitors are inscribed upon a marble slab, so a register was kept at Dodona of the people of distinction who had consulted the oracle, and all the mythical heroes of Greece were to be found in this Golden Book. Achilles and Ulysses, Orestes and Aeneas, had come to beg for advice or information from the god; nor had the gods themselves disdained to ask for guidance from the oracle: Bacchus had come to seek the reason he had lost; and Hercules, who had caused such an uproar at Delphi, had conducted himself at Dodona as an obedient son of Zeus, and had learned in return that the end of his sufferings was approaching. Many other titles to fame were inscribed in the records of the sanctuary, and in especial that of having furnished the wood of which the Argonauts built their ship.

These were brilliant records, and very necessary in order to maintain the fame and credit of the temple. Soon indeed the ancient cult of Zeus saw that of younger divinities arising by its side, and the oracle of Delphi entering into competition with that of Dodona. This was not to the advantage of the Epirot priests, who consequently spared no efforts to restore their waning popularity. They invented new ways of prophesying, similar to those of Delphi. The Pythia drank of the miraculous spring, and was seized with the sacred frenzy; the priestess of Zeus did the same, and her predictions also were in verse. Apollo had splendid games at Delphi, whose magnificence was an eyesore to the priests of Dodona; so they devised others in their turn, and celebrated the Naian games in the great theatre built at the foot of the temple. For a long time they kept the balance even, thanks above all to the uprightness with which they uttered their predictions and to the reputation for truthfulness they were able to preserve. When Delphi swerved from the straight path and the Pythia philippised, Dodona was incorruptible; and so great was the credit that it gained, that Sparta, although an hereditary adherent of Apollo, sent ambassadors to its shrine. Athens was profuse in her attentions, and all the great men of Attica, poets like Aeschylus and Sophocles, statesmen like Xenophon and Demosthenes could not sufficiently express their admiration and respect for the Epirot sanctuary.

But popular veneration is both changeful and capricious, and finally it deserted Dodona. Zeus, too, was on the decline: he had advised the Sicilian expedition, and the Athenians had sufficient reason to regret having followed his advice. No doubt he had explained after the event that the Athenians had misunderstood the oracle, and that the Sicily he meant was a small district in the neighbourhood of Dodona. Still it would have needed a robust faith to return to so clumsy a deity. Mistake followed mistake; the princes of Epirus, under the protection of Dodona, perished one after the other, to the great confusion of the oracle, and some other unfortunate incidents caused it to fall into complete discredit. A new method of consulting the deity had been invented, when an unlucky monkey chanced to upset the vessel containing the lots, and threw all the arrangements for drawing them into confusion. The oracle could not withstand the blow. It was in vain that the priests left no stone unturned. While the oracle of Ammon forwarded the water from its miraculous spring to lovers of magic spells throughout the world, the priests of Dodona submitted their sacred groves to an annual cutting, and retailed the wood in small quantities to pilgrims as a kind of domestic oracle easy to carry on a journey. It was all of no avail. Overthrown by the Aetolians, plundered by the Romans and afterwards by the Thracians, the temple of Dodona was deserted as early as the first century; and not long after, that merciless jester, Lucian, brought upon the stage the unfortunate Zeus of Dodona, forsaken by every one and whining over his forlorn condition like a high priest in an operetta.* "There was a time," says the god, "when men took me for a prophet and a physician, when I was

---

* Lucian, *Ikaromenippos*, 780-81.

everything, in short, and streets and market-places were full of the name of Zeus. Then Dodona and Pisa were brilliant and famous,—I could not see for the smoke of the sacrifices; but since Apollo set up his oracle at Delphi, and Aesculapius his surgery at Pergamum, since Thrace built a temple to Bendis, Egypt to Anubis, and Ephesus to Artemis, they are all running to these new gods ; they are holding solemn assemblies and offering hecatombs. As for a poor worn-out old god like me, they think it honour enough if once in five whole years they offer me a sacrifice at Olympia ; and so my altars have grown colder than the laws of Plato or the syllogisms of Chrysippus." To end like Calchas when one has been Jupiter is a sad fate, and this melancholy episode, which recalls the *Dieux en exil*, must not be our last impression of Dodona. In the imaginary picture-gallery described by the sophist Philostratus, one painting represented Dodona in its glory. "The wise dove with golden wings is still perching upon the prophetic oak to repeat the oracles sent by Zeus ; fillets are hung upon the tree ; for, like the Delphic tripod, it utters oracles. Here comes one to ask a question, another to offer sacrifice. He is surrounded at the moment by a band of Thebans. As to those soothsayers of Zeus whose feet are untouched by water, and who sleep on the hard ground, they are men without thought for the morrow, with no certain livelihood. Moreover they desire none, declaring that they are well-pleasing to Zeus when they thus content themselves with whatever first comes to hand. Among these priests of Zeus one is engaged in decorating the walls of the temple, another is reciting prayers ; here one is arranging the consecrated cakes, another the barley and the baskets; one cuts the victim's throat, another skins it. On this side you will recognise the priestesses of Dodona by their grave and venerable aspect; one might say that they inhale the odour of perfumes and libations. Moreover the painter has represented the smoke of the incense which envelops the scene, and even the divine voices with which it resounds. Here is a bronze statue of the nymph Echo laying her hand upon her lips; in fact, among the offerings consecrated to Zeus in the temple of Dodona, there was a basin which resounded the greater part of the day, and which was only silent when it was touched."

I do not know whether this picture ever existed, but it certainly offers a lively representation of the reality which has passed away. In the ancient world, such as we imagine it, these austere forms of mystic cenobites, these grave and gentle figures of priestesses walking in a cloud of incense, are a strange and unprecedented sight, and we must be grateful to the learned Greek whose filial piety has excavated the site of the vanished shrine, and restored to our sight the sacred spot which was the cradle of religious life in Greece.

CHAPTER IV.

EXCAVATIONS ON THE ACROPOLIS OF ATHENS.

BOOKS OF REFERENCE:
    Prof. Gardner, "New Chapters in Greek History."
    Harrison and Verrall, "Mythology and Monuments of Ancient Athens."
    Cavvadias, *Musées d'Athènes*, 1886.
    Boetticher, "Die Akropolis," Berlin, 1888.
    Ephemeris archaiologike, 1886, '87, '88.
  For sculpture:—
    E. A. Gardner, *Journ. Hell. Studies*, vol. viii.
    Studnicza, *Mitth. XI., Jahrbuch d. d. Arch. Inst.*, ii. 155; and
    Winter, *Mitth. XIII., Jahrbuch* ii.; and
    Brückner, *Mitth. XIV.*
    Lechat, *Bull. Corr. Hell.*, xii. and xiii.
  For Temple of Athena:
    Doerpfeld, *Mitth. XI.* and *XII.*
    Reproduction of statues in *Ephemeris* and in the *Antike Denkmaeler*, pl. 1, 2, 18, 19, 29, 30, and 39.

BEFORE the hour of full expansion arrives, in which art reaches its perfection, there is in all schools a time of slow preparation, during which the artist, whether painter or sculptor, is groping his way, and by patient and fruitful efforts sowing the seed which time shall ripen and gather in. Before Raphael and Michael Angelo came the early masters of Florence and Sienna, before Phidias and Praxiteles, the early Attic sculptors. It is true that the works of these old masters are sometimes very *naïf* in their inexperience; their hand is still uncertain, and sometimes fails to express their thought;

they are fettered by traditional methods, and the conventions of their school, yet this inevitable awkwardness is redeemed by precious qualities—by the sincerity of their inspiration, by the freshness of their imagination, and by the singular charm of that grace, not yet free from stiffness, which marks the works of early art. The works of these ancient schools have yet another attraction: they arose amid closely circumscribed surroundings and under certain narrowly limited influences which have made each of them original; whereas in a period of full artistic development there is a free exchange of methods and ideas, early schools are as it were thrown back upon themselves, and draw from their own stores the inspirations which they render on marble or on canvas. The Italian cities of the Middle Ages—Sienna, Assisi, Perugia—each strike a different note in the great harmony of art, and thus they leave on the mind of a visitor a distinct impression which charms him by its very unity; and in the early Attic school as it has been revealed to us by the recent excavations on the Acropolis the case was almost the same. Until the last few years the old school of Attic sculptors was little more than a name; all our knowledge of it was derived from a few passages in the works of ancient writers, and a few artists' signatures; its works had almost wholly disappeared. All we knew was that artistic development was slower in Attica than in other parts of Greece. Attic art scarcely existed at the time when the schools of Samos and Chios were flourishing, and those of Sparta and Sicyon were already springing up. Towards the middle of the sixth century we meet with one name—that of the sculptor

Endoeus, who according to the legend was a pupil of the famous Daedalus, and this name, with a few others, represented all we knew of the first school of Attic sculpture. Later on, at the end of the sixth century or the commencement of the fifth, we caught a glimpse of a second group of artists, Antenor, Critius, Nesiotes, Hegesias; and it was believed that in a celebrated group now in the Naples Museum we might catch a faint echo of one of the most famous works of this school—the Tyrannicides, Harmodius and Aristogiton. But in reality we pos-

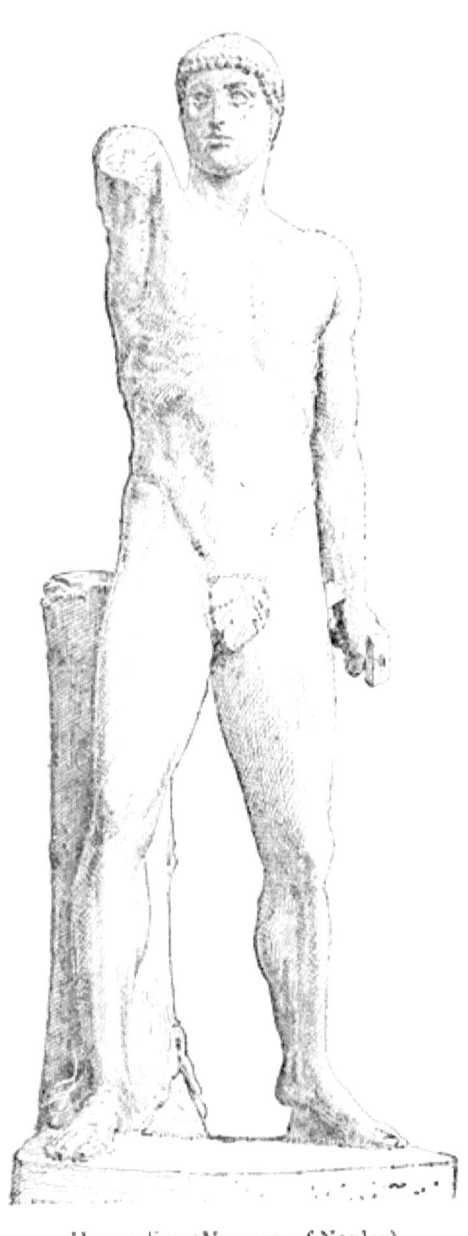

Harmodius (Museum of Naples).

sessed nothing more than the name of any of these
masters. A few sentences from ancient critics attempted
to describe their style, and from these scanty notices
we divined that it was sinewy and dry, " rigid, exactly
outlined, and marked by strain and effort." *

The recent excavations on the Acropolis have thrown
a flood of light upon the history of this early art, in a
way which is as remarkable as it was unexpected. On
the spot where the masterpieces of Phidias seemed to
have left no room for any glory but his own, on ground
which had been explored in every direction, and which
seemed to admit no hope of any further discovery, there
has been found an incomparable series of precious
monuments. Under the heap of *débris* above which
the Parthenon of Pericles rises like a statue on its
pedestal, fragments have gradually been brought to
light of all the ancient monuments which covered
the rock of the Acropolis before the Persian wars; and
the museums of Athens contain to-day a collection
of statues in marble, bronze and calcareous tufa, or
"poros," which the soil has given back to us still
brilliant in the freshness of their colouring, and adorned
with the strange and mysterious charm of early art. In
this unbroken series of monuments, which brings before
us the gradual progress of Attic art from its earliest
days, science has found invaluable data, not only for
the history of this branch of art, but for the early
history of Greek sculpture as a whole. The artist and
the archæologist have found materials for long years
of study, and the lover of the beautiful that which will
both excite his curiosity and satisfy his admiration, in

---

* Lucian, *Rhetorical Precepts*, p. 9.

these wonders of an art which is at once *naif* and scientific. We cannot be sufficiently grateful to the Greek government, which has undertaken the immense labour of these excavations, or to M. Cavvadias, the Director of Antiquities, who has brought them to a successful termination. Thanks to him, discoveries have been made on the Acropolis which undoubtedly form the most memorable archæological achievement of this century, if not of all time.

The Athena of Endœus (Acropolis Museum).

I.

As early as the year 1863, several discoveries made by chance had raised a suspicion of the riches which were still buried under the masses of *débris* accumulated on the Acropolis. At the foot of the northern slope of the rock below the Erechtheum had been discovered a seated statue of Athena, which was a curious relic of archaic art; and in the following year a statue of a man carrying a calf on his shoulders, which was soon known under the name of the Moschophorus, was found on the east of the Acropolis. At the same time other fragments of statues, and especially a fine head of the goddess Athena, were found in the great mass of earth which extends below the eastern front of the Parthenon. A single glance was sufficient to show that these figures had nothing in common with the art of Phidias; but at the moment other matters claimed the attention of the Greeks, and no attempt was made to follow up the vein which chance had opened.

It was not until 1882 that it was resolved to attack the great embankment, thirty-three feet high, which supports the Parthenon on its eastern side, and then a number of interesting discoveries were at once made. There were found, among many other things, a marble bas-relief of very early style, representing Athena seated and receiving the homage of her worshippers, also statues of women in long robes enriched with bright-coloured ornaments, among them a marble bust of exquisite freshness and grace; and above all, a multitude of scattered fragments which, when put together with admirable patience, make it possible to reconstruct several fine decorative sculptures. Some fresh fragments belonging to the head of Athena, discovered in 1863, afforded a glimpse of the attitude of the goddess, represented as in the heat of battle; while others, again, by showing her adversaries, enabled archæologists to divine the subject of the scene. It was an episode of the struggle between the gods and the giants; and the group, as is shown by the condition of the marble, left unworked at the back, served to adorn the pediment of some very ancient temple. A still more remarkable discovery was that of fragments from another pediment, representing

Hercules fighting with the Lernaean hydra, which is undoubtedly one of the oldest remains of Attic sculpture. Both material and technique are, in fact, particularly strange and curious. Carved from a block of tufa, rather coarse in grain and grayish in colour, the figures stood out in very low relief against the background of the pediment. In order to supply the deficiencies of a still unpractised chisel, the artist called colour to his aid, and the red hue of the flesh and the black of the eyes, hair and beard throw the figures into stronger relief against the deep blue background.

At the same time, these were merely interesting archæological discoveries, of the kind which only reveal their secrets with difficulty to the few experts, not such as by their brilliancy excite the interest and strike the attention even of the uninitiated. A success of the latter kind was, however, about to crown the excavation on the Acropolis. These excavations had been resumed in November 1885 in the space between the northern wall of the Acropolis and the end of the Erechtheum, when suddenly, on February 5th and 6th, 1886, fourteen female statues of Parian marble were discovered, eight of which, by a piece of good fortune rarely vouchsafed to ancient marbles, had kept their heads upon their shoulders. The circumstances of this unexpected discovery were well fitted to stimulate curiosity. Buried from ten to fourteen feet beneath the surface, in the midst of a confused mass of *débris* in which ancient inscriptions, rough stones and fragments of buildings were heaped together, and hidden in a layer of rubbish which extended from the rock itself up to the foundations of the Acropolis-wall, these statues had preserved in their hiding-place a wonderful brilliancy of colouring. Scarcely had the marks of fire in a few cases dimmed the bright colours which adorned their faces and set off the whiteness of the marble with brilliant designs. Whence came these strange idols, so unlike the goddesses

Archaic female figure (Acropolis Museum).

of Greece? What century gave birth to these maidens with their dyed hair and painted faces, with their ornaments and jewels? What do these heads represent, with this strange expression, this mysterious and haughty smile? What train of circumstances immured them in this subterranean prison, from which the pickaxe of the excavator has just released them? While archæologists were seeking a solution of these difficult problems, each day brought new discoveries, female statues similar to the earlier ones, and bronze work of the most skilful execution, thus rendering the search more exciting, and rousing a more passionate interest in these wonderful remains of archaic art.

Although they were not all of the same date, a close relationship existed among all these monuments. In spite of differences of detail, a continuous progress was visible in them, and from the clumsy imitations of sculpture in wood up to the work of an art which was already expressive and almost perfect, all bore the impress of the old sculptors of the early schools. The attitude of all the female figures is stiff, and their pose almost hieratic. They stand erect, their arms close to their sides, and while the left hand raises the folds of their long robes, the right arm is extended and holds an apple or a pomegranate. The fore-arm, generally worked in a separate piece, is wanting in almost all these statues. They are all dressed in the same way; and this is particularly important in connection with the history of ancient dress: they wear a long, closely fitting tunic or chiton, falling in large folds to the feet, over which is a short chemisette or vest, carefully crimped and plaited, which reaches to the waist; and in the majority of cases a thick woollen mantle besides is arranged in wide full folds over the bust.

Their coiffure is truly marvellous. Like all primitive races, the Hellenes of this early age introduced numberless refinements and the most studied elegance into their toilette, both in the splendour of their dress and in the skilful and coquettish arrangement of their hair.

Among the inhabitants of the Ionian cities especially, pomatum and curling-tongs were in constant use. "They comb their curling locks," says an old poet, "before they repair to the temple of the goddess; they array themselves in splendid garments, and their tunics, white as snow, sweep the ground; their hair waves in the breeze from beneath their gilded circlets, and golden ornaments glitter on their heads." The same customs prevailed in Athens, and the fashionable youths of ancient Attica, with their long floating locks covered with jewels and dripping with perfume, almost resembled women in appearance. It is related that Theseus, when a boy, as he was once walking abroad in this rich attire, met with a very amusing misadventure. Some workmen who saw him passing in all his finery began to titter and to say, "What young girl is this, old enough to be married, who is running about the streets alone?" Theseus said nothing, but, like a true hero of romance, he unyoked the oxen from a passing cart and threw them over a neighbouring temple, thus proving conclusively that they had been mistaken in him.

This characteristic feature of archaic art is not wanting to the Acropolis statues. Their hair is arranged on their foreheads in three or four rows of little curls, one above the other, and kept in place by a metal circlet. At the back of the head the hair falls in long parallel tresses, some of which are brought forward and arranged symmetrically over the breast. Sometimes metal spirals are intertwined with the curls on the forehead, to keep them in place; sometimes simpler and lighter curls form the only ornament; while in other cases there are large golden flowers on the crown of the head, springing from an iron stalk. In all we notice the love of ornament, and the coquetry, versed in every art of pleasing, with which the archaic sculptors endowed their works. The heads of these statues are not less characteristic than their head-dress. The eye is long, sometimes narrow and oblique, always very prominent; the mouth is drawn into the typical smile, still somewhat awkward and foolish, which very early artists place upon the lips of their statues; they present, in fact, the traditional types of archaic figures, such as we shall meet with again at Delos and at Eleusis. What is most remarkable about them, however, is the brilliancy of their colouring, which makes the outlines of their features, and the splendour of their dress stand out against the whiteness of the marble. Their robes are bordered or enamelled with embroideries of various colours, with the key-pattern in green or purple and with crosses in blue or green; heavy discs of gold hang from their ears; bracelets and diadems enhance the splendour of their dress; while their reddish hair and eyeballs tinged with carmine give a strange expression, a glow of life as it were, to their faces. We need to see, if not the statues themselves, at least an exact reproduction of them, in order to understand the delicate charm of these strange faces, and the attraction of their somewhat ironical and mocking expression, in which disdain is sometimes touched with a shade of haughty benevolence; and I cannot too strongly recommend all those who have access to the *Antike Denkmaeler*, published by the German Archæological Institute, to examine the fine plate (No. 19), in which two of these marvellous statues are reproduced in all the splendour of their brilliant colouring.

There can be no doubt as to the antiquity of these works of art which have been so strangely recovered: they certainly date from the sixth century B.C. But whom do these statues represent? Are we to recognise divinities or simple mortals in these youthful figures? It was thought at first, and many still think, that we might identify them with the tutelary divinity of Athens the great goddess Athena. But, apart from the fact that the individual character of many of these heads seems to show that they are portraits, a still graver objection has been raised against this explanation. Long before the time when these statues were made, the type of Athena seems to have been already fixed; and both in the seated statue found at the foot of the Acropolis in 1863, in which archæologists have thought they recognise a work of Endoeus, as well as in a series of archaic bronze statuettes found in the course of the recent excavations, the goddess is represented with the aegis on her breast, a helmet on her head, and sometimes a shield in her hand. There is nothing of this kind in the Acropolis statues; one must seek elsewhere for the cause which gave them birth. We know that in certain ancient cults it was customary for the worshippers to place their own images in the temple of the god, that they might be always before his eyes and merit his constant favour by a perpetual act of adoration: the greater part of the statues found by M. de Cesnola in the island of Cyprus owe their origin to this custom. It appears that the same usage existed among the Greeks; and on the Acropolis itself statues of priestesses or of private individuals have been found, set up as votive offerings around the temple,—as for example the figures of seated scribes, archaic marbles of which some fragments have been preserved, and the statue of the Moschophorus, who is undoubtedly nothing but a sacrificer. In all probability the youthful female statues on the Acropolis owe their origin to the same custom; their attitude of offering and adoration seems to confirm the theory, and it is natural to regard them as statues set up in honour of the priestesses of Athena.*

We must now explain how it is that these figures, which are almost intact, were nevertheless thrown like refuse into the midst of a heap of rubbish; and in order to do so we must recall a fact which it is easy to forget in the presence of the incomparable buildings erected on the Acropolis in the age of Pericles—namely, that long before the fifth century other buildings stood upon Athena's sacred hill. The goddess had a temple there in very early times, and Pisistratus the tyrant, who was especially devoted to her worship, adorned this primitive building with unusual splendour. Other monuments rose upon the rocky plateau of the citadel, within the narrow compass defended by the old Pelasgic wall, but there is scarcely any mention of these early temples to be found in the ancient authors; all traces of them had disappeared from the Acropolis. Between the time of Pisistratus and that of Pericles a terrible disaster had fallen upon the citadel of Athens. During the second Persian war, in 480 B.C., the Persians seized the Acropolis, and avenged upon its monuments the defeat of their fathers at Marathon. Its temples were

* [For objections to this view see Professor Gardner, *New Chapters in Greek History*, p. 250.—E.R.P.]

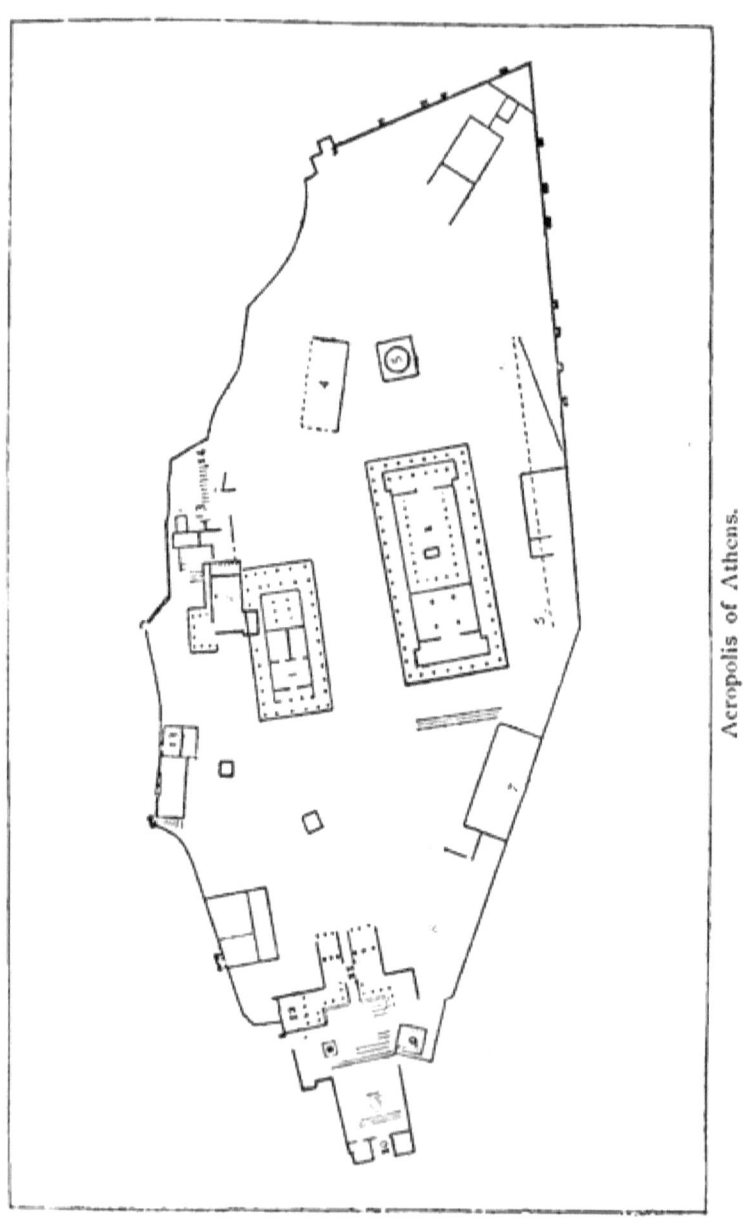

Acropolis of Athens.

1. Parthenon.
2. Erechtheum.
3. Old Temple of Athena.
4. Altar of Athena.
5. Temple of Roma.
6. Terrace Wall.
7. Chalcotheca.
8. Hieron of Artemis Brauronia.
9. Temple of Victory.
10. Beule's Gate.
11. Propylæa.
12. Pinacotheca.
13. Ancient Buildings.
14. Pelasgic Steps.

given to the flames, its statues carried away or shattered into a thousand pieces; and, as though this first catastrophe was not sufficient, in the following year the Persians returned to complete their work of destruction. When at last the victories of Salamis and Plataea enabled the Athenians to return to their ruined city, their first care was to rebuild the temples of their gods, and by the erection of stronger walls to ensure their more efficient protection for the future. It was indeed a gigantic undertaking. The Acropolis was covered with *débris*, and before they could begin to rebuild, the heaps of rubbish must be swept away. On the other hand, they wished to enlarge its area in order to give the new buildings a splendour worthy of the new city, and to protect them better by building the new walls on the outer edge of the rock; for this purpose the inequalities of the old Acropolis must be levelled, and an extensive artificial plateau constructed in its place. To complete this great work Cimon employed all the *débris*, now useless for any other purpose, of the buildings and statues which had been overthrown. What else could he have done with them? Almost all the older temples were built of calcareous tufa, and he wished to cover the new Acropolis with marble; while as for the statues, the greater part of them were lying on the ground broken into a thousand fragments, and as no sacred character attached to these offerings dedicated by the piety of the faithful, he had no scruple in employing the stone, which had lost all value through its mutilation, in the reconstruction of the citadel.

We may still see in the north wall of the Acropolis the columns, the metopes and the triglyphs of the old temple of Pisistratus, and the lower courses of the southern wall are entirely constructed with blocks of tufa from the buildings overthrown by the Persians. Everything which could be of service as building material was made use of in this way; and the remainder, that is, the dedicatory inscriptions, architectural fragments, broken pedestals, and mutilated statues, were used as rubbish to level the inequalities of the hill. This has been conclusively proved by the recent excavations. In the great heap of rubbish in which the Acropolis statues were found, several layers of unhewn stone have been observed, which correspond to the courses of the outer citadel wall. This shows us clearly in what way the building was carried on. When the circuit-wall had reached a certain height, a mass of broken pieces of stone and rubbish was shot into the empty space left between the wall and the sloping side of the rock, while over this, in order to give the workmen who were constructing the wall a solid foot-hold, was placed a layer of unhewn stones. This was repeated on the plateau thus broadened and the soil thus raised, until the day when a new foundation was ready for the new Parthenon. In this way all that had formerly adorned the Acropolis of Pisistratus still existed beneath the buildings of the fifth century, and it only needed a lucky stroke of the pickaxe to restore to us these buried treasures. From the time of the startling discoveries of February 1886, the excavations which had had so brilliant a beginning continued without interruption until January 1889. Every corner of the Acropolis was minutely explored; and the heaps of earth which covered it were cleared away down to the rock itself. For three years this marvellous vein was not exhausted; for three years the pickaxe continued to bring to light a series of remains as beautiful as the earlier products of these surprising excavations, and still more remarkable. Up to the end of 1887 marble and bronze were the materials of the finest works discovered, but in 1888 new sculptures were brought to life in calcareous tufa, very like the pediments in soft stone found in 1882, one of which represented the struggle between Hercules and the Hydra. The groups discovered in the course of these excavations seem also to be connected with the legend of Hercules; and although they were found shattered into a thousand pieces, it has been possible by patient and skilful restoration to reconstruct them in the Acropolis museum sufficiently to give us a clear conception of these ancient and interesting remains.

The first of these groups represents the struggle between Hercules and Triton, a subject found in one of the pediments of 1882, and which also figures in the bas-reliefs from Assos in the Louvre. The hero, whose head, larger than life, is painted in the most brilliant hues, kneels with his right knee on the ground, and presses the body of the monster firmly against his breast, while its hinder part stretches away in snaky coils. The other group, which undoubtedly formed a companion to the first, is very similar in design, but still more interesting. Here we see three monsters, or rather one monster with three human bodies, each with a man's head, with huge wings upon its back and with a long snaky tail, which extends in winding coils for a length of about six feet. It probably represents

the monster Typhon, defeated by Hercules, and a
head of tufa found near it seems to have belonged to
the victorious hero. The subject of the third group
is quite different: it represents a bull brought down by
two lions, which are already beginning to devour it;
its head touches the ground, its open mouth utters
its last bellowing, whilst the lions, laying their mighty
paws on the fallen beast, are
tearing it until its blood flows.

Poros head of Typhon
(Acropolis Museum).

In order to give an exact
idea of these interesting works,
our imperfect description should
be accompanied by a faithful
reproduction of the sculptures
themselves, and all who have
access to the *Antike Denkmaeler*
of the German Archæological
Institute will do well to consult
the fine plate (No. 30), which
gives one of the heads of
Typhon, that already famous,
from its brilliant colouring, under the name of Bluebeard.
One of the most noteworthy points, indeed, in these
old statues is the strange and almost violent colours
with which they are painted. Every part of the
body is covered with brilliant hues. The beard and
hair are blue, the eyeballs green; and in a little hole
which represents the pupil a sort of black enamel
lights up the face with a gleam of life. The ears,
lips and cheeks are coloured red, and the exposed
parts of the body are painted a light red, almost rose
colour, in imitation of the natural colour of the flesh.

The serpentine coils in which these monstrous figures end are painted in wide parallel bands, alternately red and blue; the same colours are found in the wings of Typhon, and make the minutely worked details of the feathers stand out clearly. The same realistic tendency is evident in the group of animals. The bodies of the lions are light red, and contrast strongly with their bright red manes; the bull is blue, with large red spots here and there, where the blood is escaping through open wounds. The colouring of the head is especially remarkable, and thanks to its wonderfully good preservation, the combined effect is as strange as it is powerful. "It would be rash to attempt to analyse the qualities of groups in so fragmentary a state. Yet in the better preserved portions we find a powerful chisel and a practised hand; they are the work of a sculptor who is still bound by certain artistic conventions, but is already free from all clumsiness."* In the Typhon group, the three bodies placed one behind the other are not the work of an incompetent artist; in the Hercules, although the muscles of the right leg are in almost exaggerated relief, the long torso is vigorous and supple, and the bodies of the serpents, so powerfully twisted and interlaced, show a desire to be forcible which has not been fruitless. It is true that if we compare these sculptures in tufa, the workmanship of which is somewhat hasty and rough, with the graceful female figures in marble of which the Acropolis Museum contains so many, we shall not find in them the same delicacy of modelling, the same minute attention to detail, or the exquisite and studied refinement in

* Lechat, "Fouilles de l'Acropole" (*Bull. de Corr. Hell.*, xiii. 141).

ornament and attire which distinguish these delightful sculptures.

The excellence of the pedimental groups consists in their somewhat rugged strength and their rather uncouth power; for the uneven quality of the stone in which they are carved did not admit of delicate and careful workmanship, nor did the place they held in the decoration of the buildings they adorned render it necessary. It is probable that these groups, which are from ten to thirteen feet in length and not quite two feet and a half in height, served to decorate the pediment of a temple. The coils of the serpents' bodies gradually grow more slender, as though they were designed to fill the gradually diminishing space in the angle of the pediment. "As we see the favour which legends of this sort have met with among archaic sculptors, we may ask whether the cause of this favour was not the ease with which the coils of these snake-like monsters could be made to fill the triangular space of the pediment."*

At the time when these interesting monuments were found, which are contemporary with the very beginnings of Attic art, other works also were being constantly brought to light to enrich the Museum of Athens. There were very ancient bronze statuettes, like that of Apollo, the modelling of which is so good and the anatomy so perfect; or those little figures of Athena Promachos, mounted on small plinths bearing the dedicator's name, which still show the marks of the great fire kindled by the soldiers of Xerxes. There were also precious marbles—heads of youths, ex-

* Lechat, *Bull. Corr. Hell.*, xiii. 133.

quisitely graceful and delicate, and winged Victories with the bust in full front view, while the legs, in accordance with archaic convention, are seen in profile, in which we already find a strange mingling of ancient traditions and new tendencies. Most important of all, there were other female statues, with the mysterious smile and the brilliant adornment usual with these painted idols, fit subjects for the brush of a Gustave Moreau or the pen of a Pierre Loti. These new-comers have the same attitude and the same costume as their sisters, as well as the same coquetry in dress, the same care in the arrangement of their hair, and the same expression; still there are several exquisite statues among them, three of which are deserving of especial attention.

The first of these is a marvel of colouring; its chiton with broad red stripe, its upper garment of dark green bordered with purple, its mantle adorned with finely-drawn designs in key pattern, the red and green crosses which are scattered over its robes and reappear in the diadem encircling its head are of incomparable brilliancy. Beneath the warm tones of these rich and exquisitely arranged colours the marble seems to turn to living flesh, and the figures breathe out a mysterious charm. Another, of later date, in all probability one of the latest of the series, shows a skilful artist striving to produce an original work. In the slender shape, the small delicate head, and the arms extended in front of the body, we trace the master's wish to depart from precedent, to do otherwise than his predecessors: the traditional smile has become almost imperceptible, the eyes which used to smile in unison with the lips no longer slant upwards towards the temples, while the hollow cheeks have grown full and round. In works of this kind archaism draws to a close; the sculptor's personality has freed itself from the traditions of his school.

The third and last is one of the most remarkable productions of Attic art. Older than the one just described it is far superior to it in artistic merit. The modelling is exquisite, and its marvellous delicacy contrasts strangely with its technique, which still savours of convention. The eyes, in accordance with the traditions of archaic art, are narrow and oblique, and the smile still draws the lips into a grimace; but these eyes are no longer fixed and expressionless,— they are lit up with life and thought, and the smile upon the lips is no longer dry and hard; it has grown soft and gentle. The statue shows no striving after novelty, but amongst those works of archaic art in which the sculptor has dutifully followed the beaten path, this, with its "open, slightly melancholy expression," is one of the most admirable.

This is not all. By the side of these ancient sculptures, the actual remains of the buildings which formerly covered the sacred hill have reappeared from beneath the layer of *débris* heaped up by the ages, and to-day we know almost as much of the Acropolis as it was before the Persian war, as of the Acropolis in the age of Pericles. The old Pelasgic walls, which confined the original citadel within narrower limits, have been discovered, as well as the remains of the ancient royal palace which crowned the summit of the hill, as at Mycenae and at Tiryns; above all, between the Erechtheum and the Parthenon, have been brought to light the foundations of the temple of Athena, which was built before the sixth century, and enlarged and adorned by Pisistratus. It belonged to the Doric order, and after it had been surrounded by a portico under Pisistratus it had six columns in front and twelve at the side. It was not built of marble, but, with the exception of the metopes, pediments and roof, of grey limestone; and after the fashion of Greek temples, it was painted in bright colours; in its pediment Pisistratus caused to be represented the battle of Athena against the giants. The hill around it was covered with other temples, and votive pillars with painted capitals supporting a multitude of statues. It is undoubtedly difficult to discover the exact situation of each building and each monument among these ruins of the sixth century; for the Persians when they sacked the temples of the Acropolis scattered their *débris* all over the hill, and it is not unusual to find fragments of the same building or remains of the same statue at a considerable distance from one another, but we may at least arrive at the main outlines of the buildings, and learn to know the early school of Attic sculpture.

The middle of the sixth century, when Pisistratus was ruling in Athens, was an important period in Greek history. States were rising, commerce was extending and civilisation developing in all directions; while poetry and the arts were awaking and unfolding their powers. We should do wrong to look upon the old masters of this period as almost barbarous, somewhat in the same way as we long regarded the early painters of Italy as barbarians. In truth these ancient sculptors deserve to be held in high esteem. The fact that they left a name behind them, and founded a school, is already a testimony to their worth; and their works, which have been recovered in such large numbers, do not contradict this presumption in their favour. None among the early schools of Greece is better known to-day than the Attic, and none better deserves to be known: before the days of Calamis and Myron, before the time of Alcamenes and Phidias, Athens had given birth to great though unknown artists, whose works discovered on the Acropolis display, throughout an almost uninterrupted series of monuments, their patient efforts and gradual progress. It is true that they did not always find their chisels capable of faithfully reproducing the boldness of their thought, or of doing justice to their thirst for truth; but the history of their attempts is none the less one of the most curious and instructive chapters in the history of Greek art.

II.

Greek sculpture and Greek architecture both derived their early models and their first lessons from carvings and buildings in wood. Just as the marble temple still recalls, in certain of its details, the wooden beams of primitive buildings, so the first rude statues which early sculptors carved of stone and marble were inspired by the primitive figures cut out of a plank or the trunk of a tree. In later times, long after art had attained its full perfection, rude idols called *Xoana* were preserved in many temples, whose figures seemed bound up in a stiff and narrow sheath, beneath which no part of the body was visible. Their arms, glued as it were to the bust, were barely indicated by a line cut more deeply in the stone; their legs were not separated, but confined in a square sheath; the closed eyes were barely marked by a horizontal line. Such, doubtless, was that ancient statue of Athena Polias piously preserved in the Erechtheum; such also in some respects were those ancient statues found at Delos, at Eleusis, and quite recently upon the Acropolis. Nevertheless these works are by no means to be regarded as the production of a really primitive art; in the Xoanon of Athens especially the execution of certain portions displays a technique which has already made considerable advances, and the face is a very free imitation of its sacred and venerable model. These works, however, are none the less interesting: they show the remote beginnings and even the starting-point of Greek sculpture.

The Greeks symbolised the early transformations of this primitive art by a beautiful legend. They attributed a great number of these archaic works to a marvellous sculptor named Daedalus, who was the first, they said, to make statues walk and see and live. He awoke the rude stone idols from their long sleep; he opened their eyes, unbound their limbs, and freed them from the immobility to which they had been condemned by the inexperience of early artists. This is the meaning, too, of the beautiful story of Pygmalion and Galatea: the sculptor gave motion and life to the sleeping statue, and made a living figure of the rude and senseless idol.

Thanks to the patient efforts of these nameless masters, an artistic type was early formed, which henceforth every generation endeavoured to perfect. Every age, in its ardent desire for progress, introduced some modification of this type; but at the same time every master, as the docile and respectful slave of ancient traditions, limited his ambition to some slight change. "Sustained in this way by the lessons transmitted to him by the accumulated efforts of his predecessors, the artist ran no risk of losing his way, and employed all his inventive powers, his boldness, his originality, in perfecting some detail of the work of many centuries. Thanks to this prudent method, no effort remained unfruitful, and the ground once gained was never lost; progress, although it was never sudden or impetuous, was rapid because it was continuous."* There are no abrupt interruptions in this art, no retrograde movements due to individual caprice; and barely a hundred years sufficed to lead the Attic school by a slow and certain advance from its modest beginnings to its highest development.

The characteristic elements of this type can be found nowhere better than on the Acropolis; nowhere can we better penetrate the obscure history of its transformation.

Let us examine all the statues discovered in the course of these excavations: some are standing motionless, their legs stiff, their attitude solemn and hieratic; others are seated, like the Athena of Endoeus, in the stiff and awkward attitude which betrays the imitation of ancient wooden statues; but in both the artist's chief care and his first thought has been the arrangement of the drapery which envelops the figure. Archaic sculpture is not yet sufficiently experienced to represent the nude; it prefers to envelop the human form in drapery arranged with the most minute care, falling

* Perrot, *Journal des Savants*, 1887, pp. 230, 231.

in heavy folds, and broken with an intentional symmetry which is sometimes very monotonous. The hair is treated with the same care as the dress, and with a curious mixture of awkwardness and refinement. In short, every external detail of dress and ornament is rendered with the utmost nicety and care. It was the face, with its numerous planes and complicated lines, which presented the greatest difficulties to the hand of the artist; and there, where his chisel is not so much at home, where his inexperience betrays itself most plainly, the characteristic elements of the archaic type can be most clearly recognised, and the progress of the school can best be traced. The eye, almost round at first, soon lengthens into an almond shape, and slants upwards towards the temples; it is narrow, like the eyes of the far East, and as yet only slightly opened, as though blinking at the light of day; while at the same time the eyeball, which was at first very prominent, grows flatter as art gains more complete mastery over its methods. The mouth, generally with full, fleshy lips, wears an awkward, affected smile, which is characteristic, and the gradual disappearance of which corresponds to the development of art. The forehead is strongly marked, the nostrils broad and full; but the most striking points are the shape of the head, which is very round, and the modelling of the flesh, which is plump and full, and sometimes even soft and bloated in appearance. Expression is as a rule almost entirely lacking in these faces; the eye is fixed and stony, the head without character. The artist feels the want, and endeavours to supply it; but in vain he attempts to give life to the face by the smile upon its lips, and to lend it the semblance of an expression,—he is compelled to have recourse to other means which are characteristic of Archaic art. Just as the architect set off the form and the decoration of his temples by the brilliancy of his colouring, so the sculptor summoned painting to his aid, somewhat like those Italian masters of the fifteenth century who heightened the dull tints of their frescoes by the application of coloured stucco. The archaic sculptor coloured his figures to make them more lifelike; he painted eyes, hair and flesh, he marked the folds of his draperies by a touch of colour—nay, more, he called metal to his aid, and heightened the vividness of the expression and the brilliancy of the eyes by ornaments and ingenious combinations of gold and bronze. In short, just as in the modelling of the body he concealed his ignorance of anatomy by the use of drapery, so he endeavoured to atone for his incapacity to render expression by giving apparent animation to the face.

It is particularly interesting to seize the history of the successive transformations of the primitive type by the study of the monuments. In this respect no series is more instructive than that of the female figures found on the Acropolis. From the coarse and clumsy work of the primitive cultus-statue down to the almost finished elegance of the latest archaic work, they display, in an uninterrupted series, the progress of Attic sculpture; and the study is the more interesting because in these figures, so closely related to one another, it is easy to detect the successive insensible modifications of the same type. There is a long succession of transitions and of more or less rude attempts between the oldest model, the *xoanon*, with its long, narrow, closely-fitting tunic, encasing the body in a stiff sheath almost without a fold, and the elegant figures which form the last stage of this evolution ; and each of these steps marks an advance. It is true that the essential elements have not greatly changed : the attitude is the same, the costume similar ; but what a difference in the more or less happy arrangement of these almost invariable elements, and what progress ! As we pass from one to another, what an effort is visible to give more life and more expression to the face ! The figure, at first massive and heavy, becomes supple and graceful ; the attitude, originally stiff and awkward, gains in grace without losing its traditional and hieratic immobility ; the drapery grows more pliant, and the robes fall in more harmonious lines. In place of the stiff bodice or cuirass which compresses the bust of the primitive *xoanon*, instead of the carefully stiffened upper garment of early archaic works, instead of the thick heavy draperies with their monotonous and wearisome symmetry, we find new combinations originated by the continual striving after progress. The difference in the materials, which have become lighter and more pliant, makes itself felt ; and a large cloak, enfolding the bust, completes the figure and gives it balance. But the advance is most apparent in the head : the hair, which was at first a heavy mass upon the head, and which, in spite of an astonishing variety of combination, always retains something artificial and affected in its arrangement, is lighter, and is arranged with simpler elegance in natural waves ; the face, which was at first round and bloated, without precise anatomy or prominent

features, is modelled with more fidelity to nature. In
the early works of this type it is still somewhat coarse
and harsh, but soon gains suppleness and life; the
surfaces, at first roughly marked out, with strong prominence given to the bony framework, are more happily
combined; the cheek-bones, originally very prominent,
with a deep furrow hollowed out in the cheek beneath
them, are less marked; the large, staring eye grows
longer; the nose is finer and thinner; the mouth
gradually loses the smile which was even more foolish
than mysterious, and assumes a less conventional expression. These successive transformations must be
studied in the statues themselves; all we can do
here is to show the final stage in this evolution,
and the point which it reaches, by two or three fine
examples which have lost almost all traces of archaic
tradition. They are to be found in two interesting
plates (Nos. 13 and 14) of the *Musées d'Athènes*, a
publication which has unfortunately been discontinued.
One is especially noteworthy. The primitive type
reappears, it is true, in this exquisite head; but what
a transformation it has undergone! The eye has
lost its former prominence; the forehead no longer
projects; the modelling of the rather long head is
refined and delicate; the profile is charming in its
grace; the mouth, serene and serious in expression,
is animated by the grave and unconscious smile which
Calamis is said to have lent to the lips of his statues;
the bust is of the most finished elegance; the drapery
perfectly simple. The master has not in this statue
succeeded in forgetting all the traditions of his school,
but his personality is striving to free itself from the

influence of conventional teaching, and in his hands archaic art has almost attained perfection. The same effort on the part of the artist, the same mingling of old conventions and new tendencies is evident in the statues which represent winged Victories. Their

Archaic head (Acropolis Museum).

attitude is archaic, and exactly similar to that of the old statue of Archermus, found at Delos, but they show a much more profound knowledge of anatomy; their draperies are raised by more impetuous motion, and their loosened locks float more freely on their shoulders. With this same group of sculptures we must connect the fine bas-relief in the Acropolis museum, which represents a woman mounting a chariot. Here we already find the long slender shapes which

Woman in a chariot (Acropolis Museum, Athens).

involuntarily recall Florentine sculpture; and we already notice that lightness and delicacy in the execution of the drapery, and that mingling of strength and grace, which were to be the distinguishing characteristics of Attic art.

The series of male figures, which are not so fully

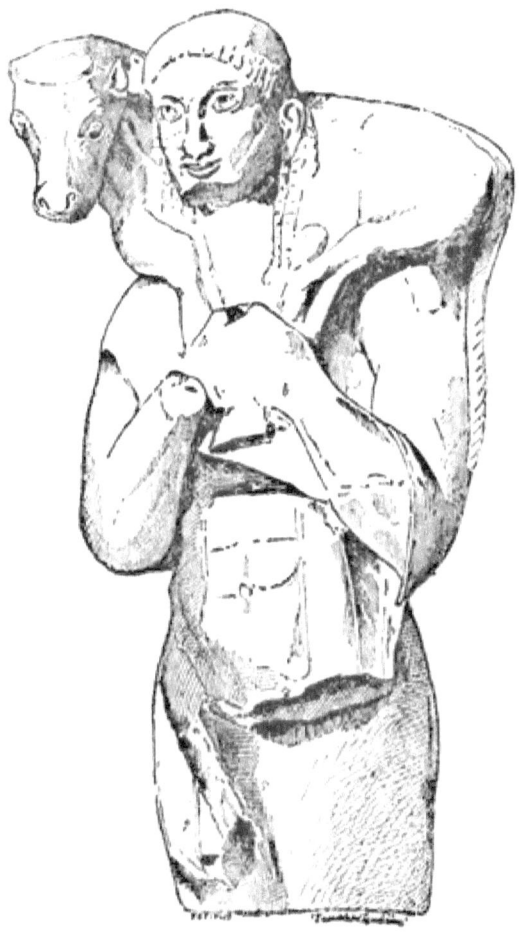

Hermes Moschophorus (Acropolis Museum).

represented on the Acropolis of Athens, are equally instructive and not less interesting. We already know the Moschophorus, with its somewhat rough, coarse contours, the strength of its modelling, the firmness of its framework, the boldness of its pose. It is one of the oldest statues found on the Acropolis, and the inscription recently discovered on its base enables us to fix its date as far back as the first half of the sixth century. With this statue should be compared several works resembling it in execution, particularly a remarkable head of an athlete of singular appearance, whose hair, arranged with the utmost care and elegance, is a masterpiece of the hairdresser's art. The build of the two heads is the same; there are the same strongly marked planes, the same prominent cheek-bones, the same large ears pressed close to the head, the same original and individual aspect which makes us feel them to be portraits, the same freedom in the composition, and, although the second of these figures is the more skilful, the same taste for dryness and precision of outline, for smooth surfaces and sharply-outlined planes. Beside these powerful works, to which we may also add the beautiful painted bas-relief known as the "Soldier of Marathon," we meet with a whole series of sculptures in which remarkable elegance and astonishing delicacy of modelling are visible. There is

Stele of Aristion (Central Museum, Athens).

the Discophorus, with his delicate elongated features, his hair arranged in plaits and held together by a kind of band, his slanting eyes, and that quaintness which is not without its charm. There are, above all, the exquisite heads of ephebi, recently discovered on the Acropolis, whose beardless faces have somewhat of the penetrating charm of Florentine work. They are

Stele of the Discophorus (Central Museum, Athens).

closely allied to the latest examples of the female type, and, like them, have retained just that trace of archaism which is needed to give them a peculiar charm. In them art has already cast off its fetters; their execution is delicate and careful almost to an extreme. One of them is especially remarkable: its chin is strongly marked, its lips compressed, its nose fine, its ears small and delicate; its eyes were once filled with glass, or with a ball of metal; its hair falls in curls over a diadem of bronze, and alone retains some traces of archaism. These works, like the latest of the female figures, bring us to the verge of the period of perfection, and may be attributed with probability to the early years of the fifth century.*

It would be not less interesting to trace in the works of this early Attic school the transformations undergone by the type of Athena, from the seated statue of Endoeus, with its stiff hieratic pose, to those little bronze figures, some of which are already so lifelike. In the former the arms still cling closely to the sides, and even the forearm is scarcely detached; a long robe, with carefully arranged folds, envelops the body, and the heavy aegis covers the breast; it possesses neither grace nor charm, and has scarcely any value except as the first of a series. Next to it we must place the Athena which decorated the pediment of the temple of Pisistratus. Here, too, all the characteristics of archaic art are still present,—the large prominent eyes, the traditional smile, the symmetrical arrangement of the hair under the helmet,—but the modelling is already far better, the cheeks are full but not bloated, the face is not without charm, and the expression is remarkably lifelike, while the attitude is full of spirit. The technique of this work is already advanced; it is true that the execution is lacking in delicacy, but delicacy would be out of place in pediment sculpture. It displays an increased breadth of treatment, combining strength and elegance; and if it is wanting in the anatomical knowledge possessed by the Aeginetan school, it is also free

* Lechat, *Bull. Corr. Hell.*, xii. 434.

from the hardness of contour which marks their work. In the bronze statuettes a further advance is made. Several of these figures are remarkable; one of them, once covered with gilding, is still a little awkward in its motions, but at least it has broken loose from the immobility of centuries, it moves, and the workmanship of its drapery is excellent. Others represent Athena Promachos, the aegis on her breast, upon her head a helmet with lofty crest, one arm raised to brandish her spear, the other stretched out before her holding the shield. The greater number of them date from the sixth century, as the accompanying inscriptions prove; yet the latest of the series are singularly lifelike, especially the one whose head resembles the Athena on the pediment of Aegina. It is true that this is not the divine form which a few years later the goddess was to receive from the chisel of Phidias; but it would be unjust to crush with the weight of that great name the early efforts of Attic art. We owe them more justice, and also more regard; for, thanks to them, during the whole of the sixth and the early part of the fifth century, Athens produced works in bronze and marble all of which bear witness to an ardent desire for progress, and many of which are noteworthy in themselves.

III.

Nevertheless, it is by no means easy to assign to the Attic school its share of the multitude of statues found on the Acropolis, and to ascertain with certainty which, among so many marbles, are from the hand of an Athenian artist. More than once in the course of the excavations, there have been found among the statues the bases on which they once stood, the light columns and pillars which raised the figure without making it heavy, and, by isolating it, made it appear more slender; more than once, too, on these bases which expand into brightly-coloured capitals, there has appeared the name of the man who dedicated the work and that of the artist who wrought it. These inscriptions show us the great number and variety of the masters who worked under the shadow of the Temple of Athena. All the great sculptors of the sixth century were represented by some work on the Acropolis—Theodorus of Samos, who was the first to work in bronze, Archermus of Chios, an authentic work by whom has been brought to light by the excavations at Delos, Alxenor of Naxos, to whose hand we owe the stele of Orchomenus, Aristion of Paros, and Aristocles of Crete, the great masters of the Aeginetan school, Callon and Onatas, and other strangers besides, such as Gorgias of Laconia and Endoeus himself, whose Ionian origin is proved by a recent inscription. It is by no means easy, at the present time, when the signatures of these artists have been violently parted from the works to which they belonged, to distinguish, among these anonymous marbles, between those which are from the hands of foreign artists and those which belong to the pure Attic tradition of Callonides and Antenor, Euthycles, and Philo, Euenor and Critius. Some, it is true, of the works themselves show unmistakable signs of foreign origin; such as the two female statues, so dull and lifeless in appearance, which belong to a different type from the other figures of the series, and offer such a striking resemblance to the Hera of Samos, now in the Louvre. In these marbles, where the absence of polychromy seems to point to the imitation of work in bronze, we may with confidence recognise the productions of the Samian school, which was the first to learn and practise the art of working in metal. Others of the Acropolis marbles offer striking analogies with the works of the Aeginetan school, or with the sculptures which adorned the pediments of Olympia; and in the presence of such a mass of evidence we cannot doubt that the Attic school was subject to varied and complex influence. But from which masters did it receive the most fruitful lessons? Whose teaching produced the greatest results? At what period did it seek their instruction? and which works, among those found on the Acropolis, have remained untouched by this foreign influence? These are delicate questions, concerning which discussion is still rife, and the jealous patriotism of the Greeks, unwilling to owe anything to others, sometimes does injury to science. Still these questions must be considered, although we cannot hope as yet to offer any but the most general solutions.

When we examine the Acropolis statues we can easily divide them into two groups. The first, and especially the youthful female figures, show a striking resemblance to the works of Ionian art. The artistic principles of the two schools are the same. The body is a mass without precise anatomy or clear rendering of the muscular system; the form is round and full, and concealed by the voluptuous amplitude and richness of the drapery. The modelling of the head is heavy: the fleshy covering, treated with loving care, hides the bony framework; the lower part of the face is soft and rounded, the lips thick and fleshy. No detail has been neglected in the careful and skilful working of the marble; the artist's chisel has delighted in endless delicacies and refinements of detail, in the intricacies of the coiffure, in the richness of the ornamentation and the coquettish arrangement of the attire. In contrast with these statues we find another group whose sculptors have been inspired by other rules. In these we find the draperies less ample, but we perceive a human body living and breathing beneath them; in the place of those profuse ornaments and that luxurious dress, simpler garments, fitting more closely to the body, allow its outlines to become visible, and it has been studied with a more watchful eye and drawn with a firmer hand.

The heads are more firmly built and less fleshy, the features are more marked, the cheek-bones more prominent, the lips less full and more closely compressed. These statues are less graceful and elegant than those of the other group, but they show more strength and vigour; the actual workmanship is less minute and skilful, but the hand that guided the chisel was more powerful and not less practised. Compare the Moschophorus, for example, with one of the female statues of the Acropolis, and you will perceive at a glance the essential difference between the two series of monuments. It is not only that the technique is not the same—there is a radical difference in their principles. One of these figures comes from Asia, and has passed from Ionia into Attica by way of the islands, through the medium of the sculptors of the school of

Chios; the other and older figure is connected with the very beginnings of Attic art.

It is in the pediment sculptures of tufa or "poros," and in statues like the Moschophorus, cut from a block of marble of Hymettus, that we must look for the earliest efforts of Athenian masters. In these works, the most advanced of which undoubtedly date from the first half of the sixth century, we already trace the names of artists rendered skilful by long practice; it is true that in the more ancient, as also in some of the pediment sculptures in tufa, the shapes are coarse and clumsy, and the attitudes awkward and ungraceful. It is also true that, in order to hide their feeble workmanship and imperfect modelling, as well as to conceal the defects in the stone (grayish tufa with occasional black spots), the artist has made excessive use of colour. Nevertheless, in the large pediments we can already trace a skilled hand and a powerful chisel. Later on, when tufa has given place to marble, although the technique is the same, and the methods formerly used in working in stone reappear in the treatment of the earliest marbles, the indigenous masters of early Athens produce powerful and vigorous works.

Parian marble, however, soon took the place of that drawn from the quarries of Attica, and with it a foreign influence made its way into the native school. Ionian sculptors transmitted their artistic teaching to Greece by way of the islands of the Archipelago; to the somewhat massive and rugged forms of the native sculptors they opposed the grace and elegance, the minute and delicate execution of their statues, the penetrating charm of their richly and carefully attired female figures; and the artists of Chios, after they had gained a footing in Delos, soon invaded Attica as well.

It was they whom Pisistratus commissioned to decorate the temple he was dedicating to Athena; it was their teaching which from the middle of the sixth century moulded the sculptors of the Attic school; and from their hands or their influence sprang the female statues of the Acropolis, and those winged figures of Victory which are so evidently inspired by an authentic work of Archermus of Chios. From the plates in the *Musées d'Athènes* (Nos. 2, 3, 4, 5) we may gain some idea of these sculptures, in which the school of Chios reveals itself in carefully finished works.

Nevertheless, the Attic school did not lose its individuality under this foreign influence. By the side of what it owed to imitation we must not fail to recognise its original qualities, an ardent desire for progress, a remarkable fertility of invention, careful and attentive observation of nature. This is no modest and timid art, such as that of Egypt or Assyria, in which the traditions and technique of a school almost exclude individual initiative. The art of the Attic school is already ambitious and inventive, and as early as the sixth century it has discovered different methods of interpreting nature and of handling its tools. Now it unites surprising delicacy of modelling with a deep feeling for grace and refinement—and again it combines a boldness of attitude, a firmness of build, an admirable moderation and precision, with vigorous and incisive execution. We trace in it two distinct tendencies, which are, however, modified and transformed with the utmost freedom, so as almost from its earliest days to unite the germs of the varied qualities which were to mark the Attic school —the delicacy and grace of Calamis, Alcamenes and Praxiteles, and the spirit and strength of Critius, of Nesiotes and of Myron.

Formerly, a few scanty fragments barely afforded us a glimpse of the merits of these old masters, while to-day they are revealed to us by a number of their works, some of which bear their signatures. It has been possible to connect several of the bases found on the Acropolis with the statues which they formerly supported; and if some of these attributions are still doubtful—as, for example that which credits Theodorus of Samos with a fine head in bronze, with curling hair—one at least is certain, and of the highest importance. On one of the pedestals which have been discovered there appears the name of Antenor, the famous artist who carved the group of Harmodius and Aristogiton, which Xerxes carried off to Susa, and of which there is a copy in the Naples museum. He was one of the most distinguished sculptors of the latter part of the sixth century. It has been possible to find, among the female figures, the one which formerly stood upon this base; and to-day Antenor's statue stands in the chief room of the Acropolis museum, " precisely as it once appeared to the Athenians—the feet and legs brought close together upon its narrow pedestal, the upper part of the body broad and amply developed, the folds of the mantle daintily drawn towards the left, the himation hanging in heavy folds, graceful in spite of the stiffness of its attitude, lifelike though motionless, at once mannered and majestic." * It is needless to dwell on the importance

* Lechat, *Bull. Corr. Hell.* xii. 151.

of this discovery, which brings before our eyes an authentic work by one of the great masters of the Attic school.*

Still, among the tourists who climb the Acropolis, there may be some who fail to recognise the charm of these old sculptures. Some will say, as they leave them, as Fauvel once said of the Aeginetan marbles: " They have neither grace nor correctness—they *are hyper-antique* and that is their only merit." Nevertheless, let us take notice that this "hyper-antique" work contains in the germ all the living principles of perfection. Wait but a few years, and the women whom these old masters have sculptured somewhat stiffly, will march, graceful and supple, along the frieze of the Parthenon. They will have learned—it is but a little thing—not to droop the corners of their lips, to dress their hair more simply, to suffer their robes to fall in more graceful folds; but they will retain their rich attire, their splendid ornaments, their modest bearing, and all the grace and elegance and delicacy which the early masters had bestowed upon them.

This archaism, too, still naïve and uncouth, has a singular charm for a highly civilised and somewhat *blasé* period like our own. In this sceptical age we take delight in the frescoes which reveal the sincere faith and candid piety of a Fra Angelico; in this period of exaggerated refinement we gladly turn to the studied and intentional simplicity which a Puvis de Chavannes has learned from the old masters. The old sculptors of

* (The author of this discovery, M. Diehl wishes to make known, was M. Franz Studniczka. Doubt is, however, cast upon this view by Mr. E. A. Gardner, *Jour. Hell. Studies*, April, 1890.—E. R. P.)

Attica charm us in the same way. When Rome grew *blasé*, and wearied of classical sculpture, the archaic schools of Greece won its favour; the collectors of the first and second centuries eagerly sought after the works of those early artists, and the sculptors of the day made them their models. We do not ask as much as that; but at least the place it merits must be given to-day to the Athens of Solon and Pisistratus as well as to the Athens of Pericles; and we must cease to consider as barbarians those masters whose hand was still somewhat unskilful, but whose talent was so real and genuine, and whose sculptures adorned the earliest buildings raised upon the Acropolis of Athens.

CHAPTER V.

THE EXCAVATIONS AT DELOS (1873—1888).

BOOKS OF REFERENCE:
  M. Homolle, various articles in the *Bull. de Corr. Hell.*
  Comptes des Hiéropes, 1882.
  Inscriptions Archaïques, 1874.
  La Confédération des Cyclades, 1880.
  L'Amphictionie Attico-délienne, 1884.
  Les Romaines à Délos, 1884.
    For the statues see 1879, 1880, 1881, 1885, and 1889; also articles by M. Hauvette (1882 and 1883); by M. Reinach (1883, 1884, and 1889); by M. Paris, 1884, 1885, 1889, and others.
  De Antiquissimis Dianæ Simulacris. Paris, 1885.
  Les Archives de l'Intendance Sacrée de Délos. Paris, 1886.
  Les Statues de Diane à Délos (Perrot, *Journal des Savants*, 1887).
  Prof. Jebb, "Delos," *Journ. Hell. Studies*, vol. i.

IN the midst of the Archipelago, almost in the centre of the Cyclades, lies a little island, or rather rock, one of the most bare, arid, and desolate that is to be found in those seas—and this is saying a great deal. It is the island of Delos. Lying beyond all the principal trade routes, only connected indirectly by a single weekly steamer with Syra, the commercial centre of the Cyclades, Delos remains unknown to the tourists who travel in crowds, " personally conducted " from Athens to Constantinople, from Constantinople to Damascus and Jerusalem. Even the cultured traveller, on his way from Syra to Asia Minor, hardly notices, as he passes, the steep and lofty summit of Cynthus, standing out against the sky, and overshadowing the whole island; in order to induce him to visit Delos itself, he must possess more ardent curiosity and more scientific tastes; he must also have, together with a passion for antiquity and a love of ruins, the courage to endure many privations. At first sight, Delos appears but unattractive; on the bare rock there is not a single tree, and only a few traces of a withered and stunted vegetation. Formerly, it is said, the island possessed a river, but no trace of it is now to be found; and at the present time the waters of Delos are limited to a brackish pond, once the sacred lake. As to inhabitants, there are none: not a village, not a hamlet recalls the existence of men; only here and there empty huts are to be seen, where each year the people of the neighbouring island of Mycono come for a few days, for the sowing or the reaping, and farther on, since the archæological discoveries have drawn the attention of the Greek Government to this barren rock, a little house for the keeper of the antiquities, who is the only inhabitant of Delos. It adds to the charm of the excursion that at Delos, as in the lion's den, it is much easier to enter than to leave. The surrounding seas are often very stormy; the wind blows furiously in the narrow channel separating Delos from Mycono, and renders navigation difficult and sometimes dangerous for the little boats which make the passage. Thus one runs the risk of remaining, sometimes for days together, weather-bound in Delos; and, as may be imagined, too prolonged a stay in this desolate island is by no means agreeable. But then, it will be said, why go to Delos at all? We go there to study the remains of its former splendour and prosperity, and to seek carefully for the scattered memories of a glorious past.

I.

To-day Delos is a deserted land. Formerly none was more rich, more famous, more venerated. Around it clustered the oldest legends of Greece, and its early history was as brilliant as its name (Delos signifies bright, shining)—as brilliant as the god of day who was born there. Leto, like many other mortals, had won the love of Zeus; and, like many other mortals, she had thereby incurred the jealous hatred of Hera. Pursued by the implacable goddess, she wandered through the world, seeking in vain a refuge where she might be delivered; Delos alone was hospitable. There at the foot of the sacred palm tree, "on the banks of the Inopus, in the golden sunshine, in the midst of perfumes exhaled on all sides," Apollo and Artemis were born.

Hence arose all the good fortune of Delos. Apollo, as the legend relates, raised an altar to himself with the horns of the goats slain by his arrows, and left it to the people of Delos as a token of his gratitude, as a pledge, too, of the great destinies reserved for his favourite island. To this primitive altar all the heroes of Hellenic mythology came in turn, to do homage to the god; on the sacred island whose inviolable neutrality offered a safe refuge to all, there landed in succession the warriors of Troy as well as those of Greece, Ulysses as well as Aeneas, the primitive kings of Attica, Theseus and Erisychthon, and in their train the inhabitants of the neighbouring islands, which, as the poets sang, were ranged around Delos like a festive garland or a chorus of worshippers prostrate before the god. After these came every Hellenic tribe, the Ionians of Greece and Asia, as well as the Dorians of Crete or the Peloponnesus; in short, all who venerated Apollo—that is, all who were Greek.* More than ten centuries before the Christian era, Delos and its sanctuary were already famous, and during ten centuries, thanks to the god whose temple it contained, Delos was one of the chief religious centres of the Hellenic world. All that Delos was, all that she possessed—her riches, her security, her poetic fame, her moral power—she owed entirely to Apollo.

"But, as mundane interests always mingle even with the most sincere religion, these men, whom their faith assembled round the sanctuary, early began to trade there; piety was associated with commerce, and a fair became an integral part of the festival of the god. It is so in all countries and in all ages; everywhere the great fairs coincide with the great religious festivals.† Delos was also an admirable commercial centre; it is situated in the very heart of the Cyclades, equally distant from Greece and Asia; placed between Rhodes and Crete on the south, and Chios and Lesbos on the north, it forms a central point in the Greek world towards which all the great trade routes converge, those from Syria and Egypt as well as those from Sicily and Italy, and also from the distant shores of the Black Sea, where according to the legend, the worship of the Tauric Artemis had already been introduced from Delos.

* Homolle, *L'Ile de Delos*, pp. 20-21.
† Ibid., p. 22.

A shrine and a commercial centre—such was Delos throughout antiquity; her whole destiny is explained by her religious traditions and her geographical situation; on her shores trade grew up in the shadow of the temple, and the temple was enriched by the aid of commercial interests. It is this double character which gives to Delos its originality: it was this fruitful union of the commercial town and the holy city which made its glory and prosperity.

We understand now how it was that the Delos of antiquity was populous, rich and fertile. Around the temple and the harbour clustered a numerous population, continually enriched by the profits of commerce and by the pious offerings which the whole world sent to Apollo. In the harbour there was a constant succession of vessels from every land, in the streets an uninterrupted line of sacred processions, before the temple a continual sound of feasting and worship. "There," said the old Homeric poet, in a hymn to Apollo of Delos, "there are assembled the long-robed Ionians with their children and their chaste wives; they wrestle, dance, and sing in memory of Apollo. Whoever should behold them then would pronounce them immortal, and unassailable by old age; so much grace would he see everywhere, and so delightful would it be to contemplate the men, the women with their beautiful girdles, the swift vessels, and the riches of all kinds. Near them are the young girls of Delos, priestesses of the far-darting god, who celebrate in their songs, Apollo, Leto and Artemis, the memory of ancient heroes and heroines, and charm the hearts of mortals by their hymns."

Once every five years, in the spring, a solemn festival recalled the anniversary of the birth of the god. The maidens of Delos, wearing their richest attire, and crowned with flowers, united in joyous chorus around the altar and represented in sacred dances the story of the birth of Apollo. Others, with garlands of flowers in their hands, went to hang them on the ancient statue of the goddess which Theseus had, according to tradition, brought from Crete to Delos. From all parts of Greece, from the islands, and from Asia, solemn embassies, sacred theoriai, landed in the harbour. The most brilliant was that of the Athenians who were long the suzerains of the island. Each year a state vessel, the Paralian galley, conveyed the sacred embassy to Delos; the crew was composed of free men, the vessel decked with flowers. At the moment of its departure the whole town was purified; the priests of Apollo bestowed on the galley a solemn benediction, and the law forbade that the purified town should be defiled by any sentence of death until the return of the vessel. The members of the embassy were chosen from the chief families of the city, and they were accompanied by choruses of young men and maidens, who were to chant the sacred hymns in honour of Apollo, and perform around the altar of the Horns, one of the marvels of Delos and of the world, an ancient and solemn dance—the *geranos*. The day of the arrival of these theoriai was a festival in Delos. Amid the acclamations of an enthusiastic crowd, the embassy disembarked in the harbour, and such was the joy and impatience of the people, that sometimes its members had not even time to don their robes of ceremony and to crown themselves with flowers. Thus more than once the sacred deputation, instead of landing immediately on the holy island, went to make their preparations in the little island opposite Delos; this was done in particular by Nicias, an Athenian general of the fifth century, and one of the most pious men of the time, who was thus able to prepare a magnificent spectacle for the Delians. He had brought from Athens in separate pieces on his ship, a wooden bridge adorned with gilding, painting and tapestry; this, during the night, he caused to be thrown across the narrow channel separating the island from the port, and in the morning the people of Delos had an incomparable spectacle before their eyes. Over the bridge wound the sacred procession of the Athenians, with its splendidly dressed musicians, its choruses chanting the sacred hymns, its rich offerings destined for the god; received at the end of the bridge by the official charged with the reception of these pious embassies, it pursued its way to the temple, there to present its offerings and prayers, and to pour out on the altar the blood of its hecatombs. During the rest of the day feasts were provided for the people, games and contests filled the island with the sounds of rejoicing. Such was the reputation of this venerated sanctuary that during the Persian wars the iconoclastic Persians, who cruelly ravaged the temples of other Greek gods, respected, by a special order from the great king, the dwelling of the Delian Apollo. Thus early was affirmed that sacred and inviolable character of the island which was destined to do so much for its prosperity.

When the gods of Paganism departed, Delos, which

lived only by their worship, fell. The homage of the world was paid to other altars, and ruins gathered around the abandoned sanctuary. It is only in our own day that life has returned to this deserted shore; once more, as in ancient times, Athens sends each year a deputation to the temple of Apollo, but not, as formerly, to guide the motions of sacred processions by the sound of chants, or to light the fires of sacrifice upon the altar. The altars have been overthrown for eighteen hundred years; the marbles, which a traveller in the fifteenth century still counted by hundreds on the strand of Delos, have become the spoil of sailors in want of ballast, or of the lime burners who have calcined in their kilns so many ancient masterpieces. Yet it is still Artemis and Apollo whom these modern pilgrims come to honour; what they desire is to find among the heaps of ruins which cover the sacred island the plan of the temple and the buildings which surrounded it, of the warehouses and the port, as well as the inscriptions engraved in marble or bronze, which relate the history of this famous sanctuary. This ambition has been gratified, and the excavations at Delos have yielded marvellous results; they have revealed the hitherto unknown history of this temple, where all the great events which agitated the Hellenic world have left their trace, where all the successive masters of the eastern basin of the Mediterranean have taken care to engrave their names and to raise monuments of their power; they have revealed in wonderful detail the administration of the sanctuary and the rites of its worship; by the discovery of curious statues they have thrown light on the development of Greek art from the seventh to the fourth century ; lastly, they have made it possible to restore the ground plan of the buildings and of the splendid assemblage of temples which so many generations built, maintained, and adorned. The honour of these important discoveries belongs entirely to France and to the French school of archæology at Athens, which quietly and with indefatigable perseverance has for long years carried on at Delos those excavations whose success has surpassed their brightest hopes.

The attention of the French school at Athens was early directed to Delos, and several papers were written on the history of the sacred island. This preliminary work, however, could yield but little result ; the soil mus be examined if it was to yield any important information, and among the extensive ruins which covered the shores of Delos it was not easy to decide where to begin the excavations. It was M. Lebègue who was entrusted in 1873 with the honour and the responsibility. His researches were directed to the summit of Mount Cynthus ; he discovered there a curious primitive shrine and two temples, but absorbed by the study and exploration of this narrow platform, led astray also by the astronomical theories of M. Burnouf, he had neither the leisure nor the means to occupy himself with the buildings near the harbour. After his departure, the Archæological Society at Athens continued and completed his excavations, and discovered also on the slope of the mountain the shrines of certain foreign deities— Serapis, Isis, and Anubis. However, it was only in 1877 that extensive and fruitful researches were begun, when Albert Dumont decided to resume the exploration

of the island and confided the oversight of the work to
M. Homolle.

M. Homolle perceived at once that these new explorations must be carried on after a different plan.
Every ancient town has a special character which
decides the nature of the exploration to be pursued
there, and, so to speak, determines the point of attack
to be chosen. Delos had been, above all, a sanctuary
and a commercial centre; it was the holy city and the
trading town which must be the principal objects of
search. All along the western coast of Delos, too,
a long breakwater was still to be seen; on the shore
the pillars were visible of those vast warehouses where
the products of Europe, Asia, and Africa had been
heaped up; and by means of well-directed excavations it might be possible to recover the plan of
the harbour, the quays, and the docks; to reconstruct the trading quarters, and restore the living
image of one of the great commercial cities of the
ancient Mediterranean. Unfortunately this task was
long, costly, and difficult, and by undertaking it they
ran the risk of once more sacrificing to a vain curiosity
the most essential thing—that is, the temple of Apollo.
It was, besides, comparatively easy to find the site of
the temple: it certainly stood on the only plain in the
island, on the only spot which was suited to the construction of large buildings, between the foot of Mount
Cynthus and the western shore of Delos; and on the
plain around it rose all those famous edifices which
former generations had accumulated in the sacred island.
But among the heaps of ruins it was not easy to find
one's way; on the plain nothing appeared but heaps of

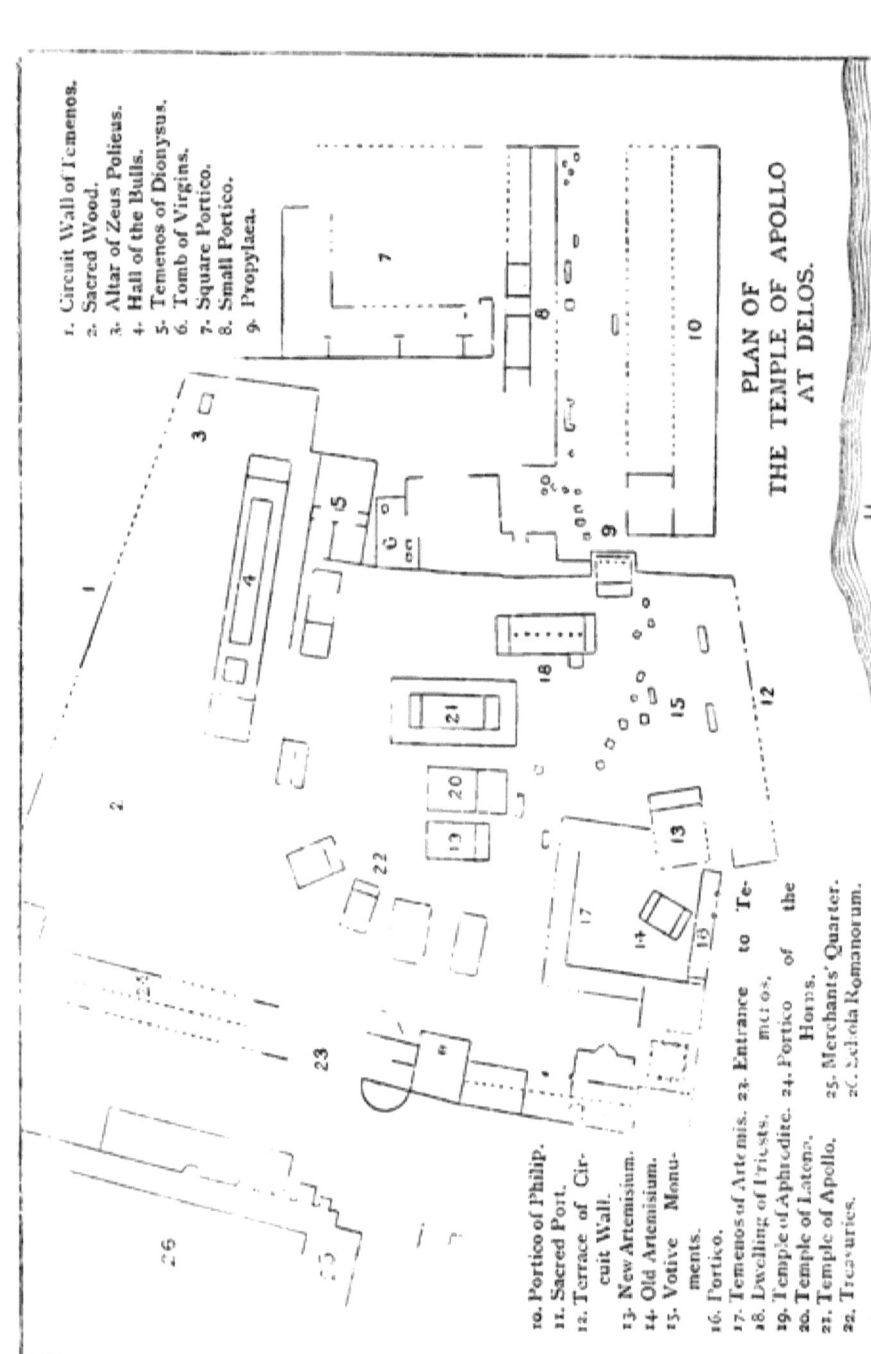

PLAN OF THE TEMPLE OF APOLLO AT DELOS.

1. Circuit Wall of Temenos.
2. Sacred Wood.
3. Altar of Zeus Polieus.
4. Hall of the Bulls.
5. Temenos of Dionysus.
6. Tomb of Virgins.
7. Square Portico.
8. Small Portico.
9. Propylaea.
10. Portico of Philip.
11. Sacred Port.
12. Terrace of Circuit Wall.
13. New Artemisium.
14. Old Artemisium.
15. Votive Monuments.
16. Portico.
17. Temenos of Artemis.
18. Dwelling of Priests.
19. Temple of Aphrodite.
20. Temple of Latona.
21. Temple of Apollo.
22. Treasuries.
23. Entrance to Temenos.
24. Portico of the Horns.
25. Merchants' Quarter.
26. Schola Romanorum.

rubbish, which here and there showed the indistinct outlines of buildings; not a single structure rose above the ground, a few stones only seemed to have remained in their places, and everywhere a layer of earth of greater or less thickness, sometimes slightly undulating, covered and concealed the monuments. One could, at most, vaguely distinguish under the heaps of ruins a few groups of buildings scattered around the sacred lake or along the western shore; towards the middle of the plain there was a vast parallelogram, before and behind which two parallel lines of ruins seemed to indicate two porticoes; to the left a vast rectangular enclosure was to be seen, with here and there the bases of statues, one of which formerly supported the Colossus, raised by the people of Naxos in honour of Apollo.*

To crown their difficulties, there was no ancient description to serve as a guide to this labyrinth of monuments. Pausanias did not visit Delos, and as to the monograph which the Delian Semus devoted to the antiquities of his native land, and in which he described in eight books the pious legends of the island, the origin and rites of the different cults, the arrangement of the temples, the monuments and votive offerings—this precious book, which would have left us little to learn as to the organisation of the shrine and the administration of the sacred treasure, has been hopelessly lost. Lost also are the treatise of Aristotle on the Republic of the Delians, and the greater part of those epic or lyric poems of which the hymn of Callimachus to Apollo

---

* Compare Homolle, " Les Fouilles de Délos" (*Mon. Grecs*, 1878, p. 26).

may give some idea, and in which poets like Demoteles of Andros, Amphicles of Rhenaea, Ariston of Phocaea, with others as well, named in the inscriptions, had celebrated the religious traditions and the local legends of Delos. All that antiquity tells us of the topography of the island may be summed up in a few lines, and those scattered and confused indications were too vague to permit even of conjecture. This important undertaking was therefore entered upon somewhat blindly, and in spite of good reasons for hope much anxiety was still felt as to its ultimate success. The catastrophe in which the prosperity of Delos perished justified all fears. In the year 87 B.C., Menophanes, a general of Mithridates, had plundered the temples, thrown the statues into the sea, and massacred or enslaved the inhabitants; everything had been overthrown, broken, scattered, and Delos does not appear to have recovered from this catastrophe. Later on the island was used as a quarry. The Hospitallers of St. John built several churches and a strong fortress from the ruins of its buildings; the inhabitants of the neighbouring islands came to seek among them for excellent building materials, cheap and ready to hand; the church of Tinos is entirely built of the stones from ancient buildings and the Turkish tombstones of the neighbouring islands are often carved from marbles taken from Delos. Less still was to be hoped in the case of the statues. Many had been carried off or destroyed in the course of centuries, and the explorers remembered with terror the exploits of that *provveditore* of Tinos who plundered the ruins of Delos to enrich Venice, and broke in the removal the Colossus of Apollo. All they expected to find were the architectural facts necessary for the restoration of the buildings, and, above all, inscriptions. As was the case with all ancient temples, the shrine of Apollo was at once a depository of archives, a treasury, and a kind of museum. The Delians deposited there the public documents relating to the administration of the town and of the sanctuary; foreigners sent thither copies of treaties and official documents to which they wished to assure full publicity; lastly, from all parts of the ancient world, numerous offerings were brought and consecrated there."[*] In order to find these records therefore it was first of all necessary to find the temple itself. After a few days' digging they were fortunate enough to discover its foundations; thenceforth they had reliable data to work from, and every stroke told.

From the first year the results surpassed all hopes, and the work begun has not again been interrupted. Between 1877 and 1880 M. Homolle made four campaigns at Delos; he has since returned there, in 1885 and 1888. In the intervals other members of the French school at Athens, whilst exploring with success other parts of the island, as well, have continued and completed the work of M. Homolle; and although the island is beginning to be exhausted, interesting or important discoveries are still made there each year. Delos has thus become as it were the property of the School at Athens, which, with its own modest resources alone, has during the last ten years won from Apollo's sacred island the greater part of its secrets.

To-day, thanks to these memorable discoveries, we

[*] Homolle, "Les Fouilles de Délos" (*Mon. Grecs*, 1871, p. 28).

can traverse with ease the sacred city of Delos, study its plan and re-construct its temples, visit its trading quarter, the quays and the port. Thanks to the inscriptions, of which more than fifteen hundred have been discovered, we can form an idea of the history of Delos till now unknown : and this history is not merely the dry and monotonous chronicle of a celebrated shrine; during many centuries Delos was a centre of religious and commercial activity, and its annals are singularly instructive in all that relates to the economic history of the ancient world. Thanks also to the inscriptions, we may learn to know the administration of an ancient temple, and the manner in which the property of the god was managed and turned to account; lastly, thanks to the statues which have been discovered, we obtain new and interesting information as to the history of ancient art. It is true that very few objects in gold or silver have been found, and not one monument in that Delian bronze so famous in the Greek and Roman world; everything valuable that the temple contained disappeared when it was sacked, and in spite of the marvellous legends current in the islands of the Archipelago as to the riches of Delos, no treasure has been found there; history has won more spoil in these excavations than archæology properly so called.

II.

There were two essential parts, as M. Homolle says, which made up an ancient temple—a building and a sacred enclosure. The former was the dwelling, the house of the god; the latter included the whole of the ground consecrated to the divinity—that is, the shrine with the buildings belonging to it, and all the surrounding space. This consecrated ground was surrounded by a wall, separated by landmarks from the unconsecrated territory in such a way as to form as it were a sacred city, which was called the *peribolos* or *temenos* of the temple. It is this sacred enclosure which we must consider first, in order to determine its limits and describe its monuments.

Let us imagine that, like the ancient traveller, we are entering the harbour of Delos in the distant days of its prosperity. Before us stretches a high wall overlooking the sea, a sort of terrace laid out in front of the temple, which forms the western limit of the *temenos*. On the right a wall of granite, now almost entirely destroyed, rises perpendicularly from the shore and the harbour, and runs eastward; behind the temple rises another wall exactly similar, built of finely-dressed stone and perfectly preserved; lastly, to the left, that is to say on the north side of the *peribolos* stretched a series of porticoes whose inner wall served to bound the *temenos*. Such was the enclosure surrounding the sacred city. To the north and south of *peribolos* along the shore the trading quarter was built, with its warehouses, its docks and the offices of its great trading companies, while higher than the temple on the slopes of Mount Cynthus rose a succession of edifices, houses and temples, whose white walls overtopped the tall trees of the sacred wood.

Such was the general aspect of Delos to the traveller. The landing-place was, as it seems, on the left, at the boundary of the two cities, in a large and richly decorated open space adorned with statues, *exedrai*, and a portico. Bounded on the east by the sacred buildings, it had a wide opening to the left in the direction of the trading quarter. Three streets went off from it: one to the north, towards the agora; a second to the east which led along the walls of the temple to the northern gates of the shrine, a grand, monumental portico opening towards the trading city; lastly, a third road turned towards the south, and led to a secondary entrance to the *peribolos*, constructed behind the terrace quite near the building called the *house of tufa* (Porinos Oikos) on account of the material of which it was built.

On each side of the enclosure there were gates, some of which led to the upper part of the plain, the others to the merchants' quarter; but the principal entrance was on the south side of the *temenos*, to the right of the harbour and the quays where the pilgrims landed. There rose the propylaea of the Doric order, the gift of the Athenians to Apollo: each of the great states of antiquity had thus brought to the god some architectural offering; each had been desirous of leaving in the sacred enclosure some splendid monument of its piety. From this structure the road started which led to the temple; its line is still clearly marked by the bases of the numerous statues which bordered this sacred way. Over all the vast space which extended in front of the temple, the most frequented part of the *temenos*, were in fact displayed the most beautiful works of art, the most magnificent offerings consecrated to Apollo, and the most important and famous public records, with which it was desired that all Greece should be acquainted. There, among many others, rose the famous Colossus of the Naxians, and among these monuments a wide passage was left free to form the road destined for the religious processions. The festival processions in all their pomp and splendour, advanced slowly across the open space, passed behind the temple, whose principal façade according to ancient usage turned towards the east, and along the two buildings parallel to the sanctuary of Apollo. The sacred way coming from the northern gates led to the same spot; and the two roads, thus joined, after having wound along a semicircular line of buildings facing the road, which were doubtless treasuries containing the offerings of the faithful, reached at length the open space immediately in front of the temple itself.

This building, such as the excavations of Delos have restored it to us, is not however the primitive shrine of Apollo; it is a building of the first years of the fourth century, erected by the liberality of Athens in the place of a more ancient structure. Built entirely of Parian marble, somewhat similar in its plan and dimensions to the Theseum at Athens, it was very small and, in spite of the beauty of its proportions, astonishes one a little by its insignificant size. The reason for this is that in antiquity the temple was intended exclusively for the use of the god; it was his dwelling, to which only the priests and a few magistrates had access, it was not the spot where the faithful came to offer their prayers and homage. It was outside the temple, on the altar built before it, that the sacrifices were offered and the crowd assembled to worship the deity. But in spite of its intentionally small size the building is of exquisite elegance, and the fragments of it which have been preserved allow us to restore with perfect accuracy the appearance of this vanished shrine. We can raise its Doric columns, which with intentional singularity were not fluted, we can replace in the pediment the sculptures which adorned it, and gain an exact idea of what this temple, one of the most famous of ancient Greece, really was.

It would be impossible to describe here all the buildings crowded together in the sacred enclosure. But among the principal ones must be mentioned two smaller temples of later date, which rose beside the sanctuary of Apollo. One was doubtless consecrated to Leto, the other to Aphrodite. There were also treasuries and refectories, and all around the *temenos*, especially along the north wall, immense porticoes where the priests lodged the pilgrims. These were, so to speak, great hostelries, which the god placed gratuitously at the disposal of his worshippers. In the same way in our own day, at the great festival of the Panegyria of Tinos and all over the East, around the Ottoman mosques there are apartments prepared at the foot of the sanctuary to receive the crowd of the faithful which recall the hospitable traditions of the old Hellenic sanctuary. Among these porticoes at Delos one was especially remarkable : it was that raised by Antiochus Epiphanes, King of Syria, which owing to its triglyphs adorned with heads of oxen was named the Portico of the Horns.

Opposite the temple of Apollo rose a curious structure, one of the most famous in Delos. It was a temple raised on pillars, which had for capitals a pair of kneeling bulls. Before this building, which formed a sort of narrow passage, 328 feet long, was doubtless placed the Altar of the Horns, one of the most venerated marvels of the sacred island. Here, in memory of Theseus, was danced the famous *geranos*, whose mazes symbolised the labyrinth of Crete; and it was doubtless through the long and narrow passage of the temple that this sacred dance wound on days of festival.

By the side of Apollo, his sister Artemis also had altars at Delos. In a separate *temenos*, situated to the north-west of the great enclosure, and bordered by two Ionic porticoes, there were two temples consecrated to the goddess. One, the more ancient, appears in the inventories under the name of the Temple of the Seven Statues; the other was called the New Artemisium. Lastly, yet other gods were associated with Apollo. At the south-east angle of the *peribolos* Zeus Polieus had an altar; farther off was the *temenos* of Dionysus, and in another part of the enclosure the shrine of Asclepius. Among the buildings of Delos were also mentioned the tomb of the Hyperborean virgins, and near the temple the dwelling of the priests or *neocorion*.

Such was the sacred city of Delos. On leaving this enclosure on either side, one entered the merchants' quarter, and there also the excavations undertaken have yielded important results. Without entering here upon an investigation of the commercial development of Delos, which is only one chapter in the history of the island, we must notice that the importance of this centre early led to the establishment of great trading companies. Shipowners and merchants established offices at Delos and soon organised themselves into powerful corporations; the inscriptions mention among others those of the Heracleistai of Tyre, the Poseidoniastai of Beyrout, the Hermaistai of Italy. Other merchants again had founded at Delos a kind of clubs which resembled at once an exchange and a temple, and were meeting-places for men of the same nationality. It was to the north of the *temenos*, between the temple and the sacred lake, that all these buildings stood, and the excavations have restored to us all those vast and beautiful structures which surrounded the lake, temples, porticoes and *exedrai*, magnificently decorated with mosaics and statues, among which one of the richest and most elegant is the schola of the Italians, with its great courtyard and vast warehouses filling all the basement. Near the offices of the great trading corporations and between these buildings and the shore, stood the agora—the market place, with the warehouses and docks where the goods were stored in great granite buildings two stories high. In this district however the thickness of the earth to be cleared away is so considerable that the excavations still remain unfinished.

To the south of the temple lay another market—the square portico. This was composed of a double gallery, one side of which formed a covered walk, whilst the other was lined with shops. Lastly, two other porticoes bordered the main street leading to the propylaea. The larger of the two, that on the west, formed a double gallery, open on one side to the sea and on the other to the street; this was also a gift, made to Apollo and the Delians at the end of the third century by Philip V. of Macedonia. It was certainly one of the most frequented spots, and one of the most fashionable promenades, in Delos; the street is lined with seats, and the numerous statues placed beneath the porticoes made it, in fact, one of the most beautiful parts of the town. Even to-day it is one of the most picturesque spots in the island; nowhere are the monuments better preserved, nowhere does one penetrate more thoroughly into ancient life. In the midst of the colossal remains of the portico of Philip, among the still living ruins of this wide street of Delos, one might imagine oneself in a corner of Pompeii.

In the plain to the east of the temple were numerous private dwellings, which even climbed the slope of Mount Cynthus. These were in general very handsome and elegant villas, the internal arrangements of which are not without importance for the history of Hellenic domestic architecture. A little to the north lay the theatre, and, further still, a recently discovered gymnasium, covered with curious drawings scratched by idlers on the walls of the building. Lastly, in the upper part of the town, on the platform immediately below the summit of Mount Cynthus, was the temple of the foreign gods. As the trading relations of the island were chiefly with the East, thence, in the train of the foreign merchants settled at Delos, came their Oriental deities as well. Egypt sent Serapis, Isis, Anubis; Syria introduced the worship of Aphrodite; here, as everywhere else at Delos, commerce and religion went hand in hand.

Such is the topography of Delos; in these surroundings once moved the people of the sacred island. Thanks to these excavations, a whole town has risen anew from the heaps of ruins; and so undeviating are the laws of Greek architecture, that a few fragments serve for the reconstruction of a whole building. It is not simply, therefore, the skeleton of a dead town that the explorations have given us: the pencil of the architect can retrace the exact image of the fallen buildings; and M. Nénot, the skilful colleague of M. Homolle, has been able to raise again the long-destroyed temples of Apollo's holy island. Thanks to the accuracy of this restoration, Delos lives again: the solitudes become animated, and its vanished splendour shines with its ancient brilliancy. Nor is this all: imagination, after having restored to Delos its buildings, can also, without difficulty, restore its inhabitants. It is only necessary to question the archives found in the temple, all those public records laid up under the protection of the god in the precincts of his sanctuary, all the mass of epigraphic material, decrees, dedications, complimentary inscriptions, found among the ruins: and these inscriptions will tell us with wonderful certainty, precision and fulness of information, the long and curious history of Delos and its temple.

III.

The inscriptions of Delos range from the seventh century before Christ to the first century of our era. It is true that they are unequally distributed among these eight centuries; but nevertheless they throw some light on all periods of its history. It is true that, for the distant times of the seventh and sixth centuries, for that obscure and half-legendary age when, nevertheless, Delos already held a high place in men's regard, the monuments are rare; a few statues,

a few inscriptions accompanying these offerings are the
only representatives of this primitive period; at the same
time the sources of these gifts, which come from Chios,
from Cos, from Naxos, agree with the impression
which the Homeric hymn gives us of Delos, the meeting-
place of the islanders of the Archipelago, the centre of
the worship of Apollo. When, in the fifth century,
Athenian rule was established at Delos, it might seem
that inscriptions would become more numerous. It is
not so however. In order to know the history of Delos
in the fifth and fourth centuries, we must go to Athens:
it is there that we must seek for the archives of the
temple of Apollo and the accounts of the sacred treasury.
During this period the political life of Delos was destroyed for the advantage of its suzerain, and only a very
few records relate to this time. But at the end of the
fourth century the inscriptions become more numerous.
At this moment, about 315 B.C. Delos became once more
the mistress of her own destinies, and then began the
most fortunate and brilliant period of her history. For
a hundred and fifty years, during the whole of the third
century and the first half of the second, the renown of
her sanctuary drew to her the homage of the whole
Greek world, enlarged as it now was by the conquests
of Alexander. Her influence was felt throughout the
East, and her commerce prospered surprisingly. It is
true that in 166 B.C. Delos lost her independence, and
was restored by Rome to the Athenians; but under the
protection of the Romans her progress was not arrested,
and this last epoch of her history was perhaps the most
brilliant. These three periods in the history of the
sacred island deserve to be studied with equal attention.

During the whole of her political life Delos, through her geographical position and her religious renown, never led a life apart from the rest of Greece; and all the great powers which aspired to the rule of the eastern Mediterranean tried in turn to bring under their authority, their protection or their influence, this post which in the hands of an enemy might become formidable, this shrine whose renown was so glorious, this port which offered so convenient and secure a situation for commerce.

When, immediately after the Persian wars, Athens saw the necessity of uniting all Greece in order to maintain a successful struggle against Persia, a league of all the Hellenic tribes, both of the mainland and of the islands, was formed under the presidency of the Athenians. In order to continue the war, and free the Asiatic Greeks, both money and a fleet were needed; each of the allies was called on to furnish a contribution and the common treasure was placed under the protection of Apollo at Delos. In all periods, ancient confederations were accustomed to choose for their headquarters the temple of some deity. The respect lately shown by the Persians for Delos had proved the sacred character of the island, which made it an inviolable refuge, and its central position exactly suited a league of all the Greeks. Here therefore, were deposited the funds of the confederation, and here assembled the council of the Delian confederacy. This honour cost Delos its liberty. Athens, who as president of the league had the control of its funds, once established in the island, was in no haste to leave it. Under the pretext that the treasure of the league was deposited in the sanctuary, she took upon herself the government of the temple, and placed the management of its affairs in the hands of Athenians. When, in 454 B.C., the federal treasure was removed to Athens the pretext was at an end, but not the usurpation. An appearance of political freedom was left to the Delians, they had magistrates called archontes, and some share in the management of the temple, but Athenian commissioners, the Amphictyons, were the real rulers. Delos, with its central situation, and the riches and strength it already possessed, might by its natural development, or in the hands of a skilful enemy, become a formidable rival; it was safer, therefore, to retain possession of it. Nor was Athens a lenient ruler; she treated the Delians with very little consideration. In 420 B.C., under pretence of rendering the sacred island truly worthy of Apollo, she ordered a general purification of the territory; in order not to defile the sacred soil the ancient sepulchres were removed, and restrictive regulations henceforth forbade that either births or deaths should take place on the island. No doubt there was some resistance, and this was overcome in 422 B.C. by a more radical measure, the total expulsion of all the Delians. Thus Athens crushed a possible rival, and, as she believed, secured permanent possession of the temple. The misfortunes which marked the close of the fifth century destroyed these hopes together with the power of Athens. After the victory of Sparta, the confederation was dissolved, and the separate states regained their liberty, Delos among them. Its inhabitants returned to the island, and were restored to their ancient rights. This did not last long, however, Athens soon took possession of

Delos again. In vain did the revolt of 375 B.C. show the ill-will of the subjects towards their masters, in vain did the Delians ask everywhere for help, their complaints were ineffectual, their efforts useless, their appeals unheard.

The fourth century was in every way a sad time for Delos. This can be discerned in the inventories, in which are set down the riches of the temple and the gifts made to Apollo by his worshippers. In these records, which give us an abstract of the history of Delos, no period is so poor as the fourth century. Athens, formerly so liberal towards the god, so ready to celebrate magnificent festivals, to send solemn embassies to Delos, now neglects Apollo. The great sovereigns of the time are heedless of the shrine; the names of neither Philip nor Alexander appear in the lists, which seems to prove that Delos had fallen into utter decay. Besides, if the place had been worth the trouble of seizing it, would Philip have left it to the Republic? The Athenians retained possession of it rather as a matter of pride than of profit; they remained in the island, but without being able to do anything for it, and Delos, without military or commercial importance, could no longer possess much religious influence.

After the death of Alexander, in the last troubled years of the fourth century, when the Macedonian generals, in order to gain partisans and to secure a part in the heritage of their master, vied with each other in proclaiming liberty to the Greeks, Delos escaped from the Athenians, and regained the control of the temple. Mistress of herself, respected for the antiquity of her worship, sought after for her moral influence, she

became at that moment a political power. Around her
was grouped a confederation of the neighbouring islands ;
and this league situated in the centre of the Archipelago,
became the necessary ally or *protégé* of every power
which desired to rule the Aegean Sea. Thus Delos at
that moment became of importance in Eastern affairs.
She sent and received embassies, she had a foreign
policy. It is true that among the great powers which
contested the rule of the Eastern seas or in turn succeeded
to it, between the flourishing republic of Rhodes, and
the kings of Syria, Egypt and Macedonia, the league of
which Delos was the centre was not so much an ally as
a *protégé*; a town like Delos, which had neither army
nor navy, and which depended for food on external
supplies, needed safety above all things, and this could
only be found in the protection of a great power. But
if her position was sometimes delicate and difficult
among these conflicting influences, on the other hand
she also profited by them. The development of Hellenism which was the consequence of Alexander's conquests, the new life which Greek genius infused into the
whole East, brought new worshippers from all quarters
to the gods of Hellas. Just at this point the inscriptions
allow of our determining with precision the sphere in
which the influence of Delos was exercised, and in which
its interests were concerned. Peruse the long series of
dedications inscribed on the bases of the statues. The
names, the nationality, the rank, of those who present
these offerings, the number and splendour of the
monuments dedicated, show the movement of population
kept up in the island by religion and commerce, and
inform us of the principal countries which had relations

with Delos. Study, on the other hand, the decrees voted by the senate or the people; the nationality of the persons honoured in these inscriptions show in what direction are the most active and cordial friendships of the sacred island. Take in your hand the inventories, where, together with the offerings, are inscribed the name and country of the donors, and side by side with the list of the countries towards which Delos felt herself drawn, one may make a list of the nations and kings who felt the attraction of the holy island.

First, at the end of the fourth century, comes the powerful republic of Rhodes. At this time, and throughout the third century, the Rhodians had a monopoly of maritime commerce in the East. They served as middlemen between the peoples of the eastern basin of the Mediterranean, bringing goods from the country where they were produced and selling them again to the dwellers on the shores of the Aegean. All the external trade of Egypt was in their hands; they sailed to the Black Sea to buy the wheat of the Crimea, and supplied the whole Archipelago with it. The importance of Delos to them may be imagined. It formed an admirable market and emporium; the festivals of Apollo, the fame of these rites, attracted an extraordinary crowd of pilgrims from all parts; the sacredness of the island assured its neutrality, and thus they were sure of finding there at all times a large number of customers. The Rhodians quickly saw all these advantages. The troubles of Athens, the rivalry of Alexander's generals, left them a free field; and it was doubtless at their instigation and under their protection that the league of the islands around Delos was formed. Henceforth

Rhodes sent each year embassies and rich offerings to Apollo and in return the people of Delos overwhelmed the chiefs of the great maritime republic with compliments and honours.

After Rhodes it was Egypt which, in the third century became the patron of Delos. The Ptolemies early saw the advantages to be derived from its geographical situation and its religious influence, they also saw that the control of the island was the necessary condition of the establishment of any great maritime power in the eastern Mediterranean. Therefore they rendered all possible honour to Apollo : throughout the first half of the third century the lists of offerings are full of the names of the Ptolemies ; every year solemn embassies brought presents of great value to the temple ; every year festivals were held in the island, established expressly in honour of the kings of Egypt, to which they had the privilege of giving their names. Between Alexandria and Delos there was constant interchange of courtesies ; Egyptian fleets protected the sacred island and collected for Apollo the arrears of tribute owed by the islanders, while, in return for these good offices, Delos erected statues to all the princes of Egypt, to their wives, to their sons, to the chief personages of their court—one statue to the king's physician, a second to the director of the museum, a third to the architect of the Pharos. There are endless decrees in honour of Egyptian officers, admirals or governors of provinces, who ruled the islands in the king's name, as well as decrees in honour of the vassal kings of Egypt who, following their master's example, came to do homage to Apollo. Thus, during half a century Delos was under the protection of the Ptolemies, Alexandrian money was current in the island, and an Egyptian admiral governed the Cyclades.

It was not however without a struggle that Egypt established and maintained this maritime empire. The kings of Syria and Macedonia more than once disputed her influence on the sea, and as the inscriptions of Delos show, it was often with success. Amidst these rivalries Apollo showed himself a consummate diplomatist. The constant policy of Delos was to make the neutrality of her sacred soil respected as much as possible; even in times of war she received impartially in her harbour the fleets of either belligerent, and accepted without scruple offerings from any hand. At the same time as the Ptolemies, the kings of Syria loaded Apollo with gifts, and Delos rewarded their liberality with statues. Beside the distinguished persons of the Egyptian court, complimentary inscriptions name the favourites, the ministers, the physicians, of the Syrian kings; beside the festivals in honour of Ptolemy, there were also festivals in honour of Seleucus. The case was the same with Macedonia; Delos honoured Demetrius, Antigonus or Philip with as much devotion as their rivals, she accepted the gifts of the one with as much gratitude as those of the other. Thus she assured to herself the good-will of her future masters. At the end of the third century, in fact, the protectorate of Delos passed to Macedonia, and the portico of Philip is the signal proof of the benefits which the Macedonian rule conferred on the islands.

As Macedonia had succeeded Egypt, so Rhodes succeeded Macedonia at the beginning of the second century. During the long conflicts of the third century, the constant aim of the great maritime republic had been to protect the islands of the Archipelago from every conqueror who might have threatened her own commerce. Before all things Rhodes had endeavoured to preserve the freedom of the sea, and had held the balance in turn between the states which aspired to rule the Aegean ; and thus, when Egypt and Macedonia had successively lost the supremacy, she remained the only considerable power in the East. At the head of a great commercial confederacy, not unlike the Hanseatic League of the Middle Ages, Rhodes now ruled and protected the Cyclades. Her fleets assured the safety of the seas and the neutrality of Delos, and in that unequalled market her merchants made full use of the commercial monopoly which they had secured for themselves. It mattered little to the Delians who was their master ; they were sure to profit by any rule. From all parts of the Hellenic world offerings and worshippers came to them—from Greece and Asia, from Egypt and from the islands. From Cos, Rhodes, Alexandria, solemn theoriai set out each year for Delos, and the great powers of the East could not show too much respect and devotion to Apollo. If Athens was wanting and seemed to have a grudge against the god, the Delians were easily consoled for this by the crowd of pilgrims sent to them each year by the East. In acknowledgment of their liberality Delos decreed statues and festivals to them all, and all in return covered the sacred island with magnificent monuments. One king of Syria built the Portico of the Horns ; another filled Delos with statues ; Philip of Macedon erected a portico ; and less important rulers gave in proportion. Thus the island passed from hand to hand, always honoured and renowned, she grew rich and beautiful, and her inhabitants tranquilly passed their lives in receiving and entertaining strangers, and in earning a good deal of money by fattening poultry, and acting as innkeepers and slave-dealers in turn.

Trade prospered at the same time as religion. Placed in the centre of the Cyclades, at the junction of the great routes of the ancient world, possessing an excellent port, Delos had all the advantages necessary for a commercial centre. The Delians had business relations on all sides—with Alexandria, with Byzantium, with Rhodes, with Crete. A far-seeing and skilful diplomacy regulated their friendships and their alliances; together with religious motives, more mundane reasons guided their sympathies. The need of safety led them towards Crete, a nest of robbers who must be conciliated as they could not be resisted; economic considerations drew them to Rhodes and Byzantium, the two chief seats of that eastern trade, all the traffic of which passed through the mart of Delos. "The principal article of this commerce was wheat. The question of provisions was indeed of the utmost importance to Delos, whose territory was insufficient for her support; and therefore the island had relations with all countries which produced corn, with Cyzicus and Lampsacus, with the Chersonese and Panticapaeum, with all the states which traded in wheat. But not content with importing the quantity necessary for their own consumption, the Delians were ambitious of becoming purveyors to the whole of Greece, by storing this indispensable article in a position accessible to all; from the time of the

Rhodian supremacy Delos dreamed of becoming the common market of Greece."*

However, in spite of these ambitions, " in spite of the situation and the excellence of its harbour, in spite of its privileged position as a sacred city, Delos at the beginning of the second century was still only a town of the third or fourth order." † Between Delos and the powerful cities of Greece and Asia there was no equality, she could expect nothing from them but competition and rivalry. Rhodes aspired to the monopoly of the sea. Corinth was also flourishing, and Delos, moreover, had within herself none of the resources which make a great maritime and commercial town. Her narrow territory was not fertile; it furnished neither wood for shipbuilding, nor produce for export. Delos could therefore be nothing more than an emporium, a mart dependent on others. If she was rich and luxurious in the third century, she owed it to religion more than to commerce; she owed it to that concourse of strangers whose piety enriched the temple, and who in fact made the Delians the parasites of Apollo. But circumstances were about to change; as soon as Rome appeared in the East, her intervention in a few years raised Delos to the highest rank; her victories in Macedonia and Asia had the most important consequences for the island. The holy city of the third century became a commercial town; and it is to this economic revolution, so rapid and so extraordinary, that the final splendour of Delos is due.

By the end of the third century, the Romans began to appear in the East. From this moment we meet, at

---

* Homolle, *Rapport sur une mission à Délos*, p. 36.

† Ibid., *Les Romains à Délos*.

Delos, with the first nucleus of that colony of Roman traders and capitalists, which soon became so prosperous; from this moment the political instincts of the Delians brought them into connection with the great republic, whose power was constantly on the increase. When, shortly after, a war brought the Roman squadrons into the Archipelago, the generals of the Senate soon became acquainted with the port of Delos—" a safe and convenient harbour in these often dangerous waters, an excellent post of observation in the centre of the Cyclades, a refuge doubly protected by nature and by religion.* "They found the place suitable; and this strategic position was quite naturally destined to become, in time of peace, the chosen site of the great maritime emporium of the eastern Mediterranean. In fact, whilst the West remained outside the circle of trade, Delos could only have a secondary importance, but now that a new customer, Italy, had to be taken into account, a new market was needed, where the merchandise from different regions could, halfway from their place of origin, be gathered together, and brought within reach of this important purchaser. The Romans perceived the suitability of Delos for this purpose, and in the train of the Roman generals who came to do homage to Apollo, and who exercised a political protectorate over the island, there soon arrived capitalists who established banking and commercial houses at Delos; and in a few years this colony was prosperous and influential enough to hold the first rank in the island. It was at the request of these bold and selfish traders, that the Senate took in 166 B.C. a measure which

* Homolle, *Les Romains à Délos.*

was decisive of the future of Delos. In order to crush all outside competition, and especially that of Rhodes, in order to concentrate in the island of Apollo all the trade of the East, Rome constituted Delos a free port. This was, for the Roman banking houses established there, an essential condition of their prosperity. At the same time, in order to secure more completely the Roman preponderance at Delos, the Senate expelled and proscribed the Delians *en masse*, afterwards restoring the island to its former masters, the Athenians. But, in fact, this restitution was only a diplomatic fiction, which could deceive no one; it was the Romans who arranged it all and who were to profit by it. In appearance the government was Athenian, but in reality with Athens was always associated the Roman colony, in whose hands was all the wealth, and who managed all important affairs.

Rhodes, which up till then had had a monopoly of trade, was at once ruined by these measures of the Senate; in three years the revenue from her customs fell from a million to a hundred and fifty thousand drachmae. One more piece of good fortune came to complete the prosperity of Delos; it was the destruction of Corinth in 146 B.C. Henceforth Delos had no rival, and her fortunes rose to a marvellous height. Having become the trading centre for Greece, Asia and Italy, she had a monopoly of the Eastern trade, and this monopoly was extremely lucrative. All the articles of luxury which the East produced found purchasers at Rome, and sold at high prices there; in order to supply this demand, works of art, costly stuffs, spices—all the products of Greek and Asiatic countries—were gathered together at the great festivals, or rather fairs, of Delos. Slaves abounded in this celebrated market, where in a single day more than ten thousand head of these human cattle were sold. Under the powerful protection of Rome, trade was carried on in safety, and when in 133 B.C., Asia became in her turn a Roman province, and the inexhaustible riches of that country fell into the hands of Roman capitalists, this was a new advantage for the free port of Delos.

At this point, about thirty years before the end of the second century B.C., we will stop for a moment, in order to take a more extended survey of the commercial prosperity of Delos. Here again it is from the inscriptions that we must ascertain the chief directions followed by Delian trade and the vast field over which it spread. A host of monuments erected to Eastern princes, to the sovereigns and officials of Pontus and Bithynia, of Syria and Cappadocia, of Parthia and Egypt, attest the relations existing between these kingdoms and the sacred island. "It is towards the Levant, in fact, that the commerce of Delos turned exclusively; its principal seats were Syria and Egypt, which held the first rank in the industry of the time, and were in direct communication with the producing countries of Asia, of the interior of Africa, and of the extreme East. In Egypt all the trade was concentrated in Alexandria; in Syria, twelve towns, the Ladder of the Coast, served as halting-places between the interior and Delos."* From all these countries worshippers crowded to the sacred islands; not only did princes and great men make offerings to the temple, and think

---

* Homolle, *Les Romains à Délos*, p. 110.

it an honour to have their statues there, but private persons came in large numbers. Men of every race and country—Syrians and Egyptians, Romans and Athenians—mingled at Delos, and soon were numerous enough to form rich and powerful corporations there. As early as 150 a guild of Tyrian merchants was formed under the protection of Hercules; then the merchants of Beyrout were associated under the auspices of Poseidon; later still the Apolloniastai constituted another religious and commercial association, also placed under the patronage of a national deity. The Jews, the Syrians of Hierapolis and Ascalon, were organised in the same way; the Egyptians, anxious to preserve their national customs, formed themselves into a synod, which seems to have had considerable importance; but of all these foreign colonies the richest and most influential was the Italian and Roman colony, which from the middle of the second century formed itself into a corporation under the protection of Hermes, and bore rule in Delos.

With the different nationalities came their gods. Although Apollo always nominally held the first place, piety now turned more readily to younger divinities, and most of all to those gods of the Eastern world whose worship has always been so attractive. The Hellenised East invaded Greece, and Greece made no resistance: Isis and Serapis, Anubis and Harpocrates and the Syrian Astarte, had their temples, their worshippers and their mysteries, at Delos. The importance of these rites was such that Athens thought it necessary that they should be directed by Athenian priests, and in fact numerous inscriptions attest the amount of offerings brought to these shrines. Tyrians and Sidonians, the people of Laodicea and Miletus, of Bithynia and Pontus, of Egypt and Syria, pressed to these new altars, and even the Romans did not disdain to take part in this worship.

Thus Delos assumed more and more the character of a cosmopolitan town. This is shown by the continued rise in the rent of houses, together with the continued decline in the rent of farms. The more foreigners crowded to the island, the more were lodgings sought after, and the greater part of the houses were transformed into stores and hostelries. At the same time the city was beautified on all sides. The great trading associations bought land, built temples, and erected halls and porticoes where their members could meet to discuss their common affairs, and to concert measures for influencing the market. These beautiful and extensive buildings, mute witnesses to the importance of the guilds which met in them, are situated around the sacred lake. The most remarkable among them is that which belonged to the Roman colony. It is the largest, and must have been the most magnificent of the buildings of Delos. A vast court, measuring 310 ft. by 230 ft., is surrounded by a portico behind which extends a large building divided into a multitude of rooms, which opened on other porticoes and on the neighbouring streets. The whole was built of marble; under the portico were placed seats and statues; here and there are other halls which served as places of meeting and where stood the statues of the benefactors of the guild, while the basement was used for storage; long rows of amphorae or large jars, still stand there.

Nothing can give a better idea than this vast building, of the importance of the Romans in the island and of the riches of Delos.

At the same time the harbour was improved; the breakwaters, the quays, the warehouses, were rebuilt or enlarged, and these works, whose ruins are still visible, bear witness by their grandeur to the power and prosperity of Delos.

"On the whole length of the west coast, the line of the quays can be traced; sometimes on irregular blocks, flung pell-mell into the sea, there are heaped enormous masses of granite; elsewhere the rock itself has been hewn out. An immense mole, which still forms an efficient protection for the roadstead of Delos, runs out like an arm into the sea, at the northern extremity of the island another breakwater once served to break the violence of the current and farther south a line of quiet harbours was formed by means of jetties, where goods could be loaded and unloaded. The wharves gave direct access to the warehouses which lined the coast and which were separated into two divisions by the temple of Apollo. To the north, over all the hill which divides the sacred lake from the sea, extended vast buildings which stretched along the quay, ran inland parallel to the sacred enclosure, and finally turned from south to north opposite the schola of the Italians. The long succession of these halls can still be traced to-day; they were the docks of Delos."[*] To the south the portico of Philip was used for the display of goods, and, further on other storehouses lined the coast. Soon a new building was added to these edifices; to the east

---
[*] Homolle, *Les Romains à Délos*, pp. 124-125.

of the portico of Philip the Italian colony erected the square portico, which was really a bazaar. In spite of their ruinous condition, one cannot contemplate without admiration this mass of buildings, which impress one as much by their imposing strength as by their ingenious arrangement. Delos was, in fact, the model port of the Romans, the commercial town *par excellence*, and they thought they gave the highest praise possible to Puteoli, when they called it a little Delos.

The Mithridatic War destroyed this prosperity. One of the generals of the King of Pontus sacked Delos, the temples were plundered, the statues broken, the Roman colony massacred or dispersed. Delos never recovered from this blow. It is true that when the war was over the Romans returned to the island, the ruins were repaired, and business was carried on again, but new troubles fell on Delos, the pirates did not respect her neutrality, and cruelly plundered her temples. This was the fatal blow. However, Delian commerce struggled on for some time longer, but the competition of Ostia and Puteoli completed its ruin. After the catastrophe of the year 87 B.C., Eastern traders ceased to come to Delos; soon the Romans also quitted it. At the end of the first century the island was almost deserted, and if it still retained some traces of its ancient religious glory, its commercial greatness was entirely destroyed.

"The causes of this fate, apparently so extraordinary," says M. Homolle, "of this lofty greatness, this profound and irreparable fall, are easily understood. The prosperity of the island was entirely artificial; it produced nothing; it was too small and barren to support at all a numerous population; even water was wanting there. No kind of industry was carried on in the island; with the exception of its famous bronze, not one really Delian article of commerce can be named: thus, having nothing to export, and being obliged to import everything, the town could serve only as an emporium. Its very fortunate geographical position, and its religious influence, permitted it to play this part successfully, and made it a populous and wealthy city, a great centre of the carrying trade, a general resort of the pilgrims and merchants of the Greek world; but when misfortune struck this chance population, when disaster came to disperse it, or even affected its interests injuriously, it was sufficient to cause its trade to be diverted to a more fortunate town, its worship to be abandoned, and for the whole island to fall into complete and hopeless destitution, for it had no resources in itself.

"The same causes act at all times in the same manner. The regeneration of Delos has narrowly missed following on that of Greece; when it was necessary to find a stopping place in the Aegean for the lines of steamers, Delos was the first place to be thought of. Syra was chosen; it was as well situated as Delos, but equally poor and still more barren. To-day—and less than fifty years have passed—it counts nearly thirty-five thousand inhabitants, but it would fall like Delos if the Austrian Lloyd, the Messageries Maritimes, and the other steamship companies, removed their offices elsewhere. Already the competition of the Piraeus is checking its development. Tinos has inherited the religious *rôle* played by Delos, and one of the most curious spectacles in Greece is offered by the motley crowds, speaking all the languages and wearing all the costumes of the East, which press to the festivals of the Panaghia on March 27th, and August 15th.

"In all this we see the working of invariable laws and the inevitable results of situation; a favourable geographical position, commercial interests, religious beliefs, have made Syra and Tinos as they made Delos."*

IV.

"Among the temples of antiquity there is none, not even the Parthenon, the details of whose administration can be as minutely studied and as accurately ascertained as that of the Delian Apollo." † A unique series of records, such as exist in the case of no other Grecian shrine, allow us to reconstruct with the most rigorous exactitude, and in all its curious details, this side of the religious life of antiquity. On this subject alone we possess more than six hundred fragments, representing more than one hundred and fifty inscriptions, among which those of three, four, or five hundred lines in length, are not rare. These records are scattered over the whole extent of Delian history, but they throw special light on the period of one hundred and fifty years, from 315 to 166 B.C., during which the island was independent; for this period alone there exist no fewer than four hundred and fifty inscriptions or fragments of inscriptions.

* Homolle, "*L'Ile de Délos.*" Conférence faite à Nancy.
† Ibid., "Compte des Hiéropes" (*Bull. de Corr. Hellén. VI.* 1).

These records are of two kinds: the one consists of the accounts referring to the sacred property, the other gives us the inventory of the offerings and of the contents of the temples. Let us clearly understand what an ancient sanctuary really was. Before all things a god was a great landowner. Not only did the state assign to the temple an endowment in land, but the piety of wealthy private persons continually increased the extent of this domain. Besides this, the sacred treasury received the tenth of the plunder taken in war, it also received a share of all confiscated property, lastly, presents of money frequently accompanied the gifts of land, and in proportion as the renown of the god and the fame of his worship increased, so the gifts became more numerous, and coin and bullion accumulated more and more in his coffers. Thus a fortune grew up, consisting of both real and personal property. Besides these sources of income, there were also the votive offerings, which were unproductive riches, no doubt, but were of considerable value, and the importance of which increased from year to year. Celebrated sanctuaries were almost always the recipients of costly ornaments in gold, silver, or bronze. The articles which the piety of the faithful presented to these sacred collections soon came to represent an enormous sum. We can judge of this by the descriptions of ancient authors, always eager to draw up the catalogues of these offerings, by the lists of them preserved at Olympia and at Delphi, by the large number of monographs which the learned men of antiquity devoted to these treasures, by the authentic inventories of several famous temples, such as the

Parthenon and the shrine of Hera at Samos, and by the numerous ex-votos found in the excavations of Dodona and Olympia. Thus the temple was at once a bank, a storehouse, and a museum, and it was of course necessary to manage and watch over all this wealth. It is the accounts of this management that the excavations of Delos have restored to us.

There in fact, more than elsewhere, the management of the temples held an important place. "Delos," says M. Homolle, "only existed through Apollo; she owed him everything—safety, riches, poetic glory, and that moral influence of which the greatest states felt the power and sought for the support. The Delians therefore considered religion their chief, if not their only, concern. The temples, that of Apollo above all, were the objects of their continual care ; they were carefully kept up, as the seats of their worship, as a part of the national patrimony, as the source of their greatness and prosperity. They carefully preserved the innumerable offerings which formed the adornment of the sanctuary, and which were titles of honour for their town, proofs of the power of their gods, and an important part of their riches. Another motive, more powerful still, induced them to watch over the sacred treasure. Its annual revenue, its surplus income, its legal reserve, constituted a fund all the more important because the town was small and its resources limited ; the State had recourse to it in all emergencies."* For all these reasons, spiritual and temporal, the management of the treasure of Apollo was the great business of Delos.

Four magistrates appointed yearly, whose titles varied

---

* Homolle, *Les Archives de l'Intendance Sacrée à Délos*, pp. 1-2.

at different periods, were charged with the care of the sacred treasure. Two of them usually presided over the sacrifices and religious rites; the duties of the other two were essentially administrative—they were charged with the preservation of the temples, with the administration of the landed and other property, with the management of the treasure. It was they who lent out at interest the surplus sums of money, who let the farms, who leased the taxes, who gave out contracts for work to be done. It was they who had the care of the offerings, of the things needed for worship, and of the public treasure. They were therefore surrounded at once by the greatest respect and by the keenest watchfulness. Every year, on giving up their office, they had to render an account of their stewardship, to show that no property was missing, to justify their employment of the sums which had been committed to them. The handing over of the treasure by these officers, the hieropoioi to their successors was a solemn ceremony surrounded by great pomp; and the explanation is simple, for the public credit itself was at stake. In the presence of the senate and the magistrates the hieropoioi first rendered an account of the finances, and delivered up the cash in hand; they then proceeded to hand over the treasures themselves. With this object a general verification of the treasure was made; the separate articles were passed in review one by one, weighed, and described minutely, after which they were delivered by the departing trustees into the hands of their successors. The making of this inventory served as a release for the former, and as the formal taking over of their charge by the latter. Lastly, the lists of the offerings were published and preserved at great expense; the stelæ or slabs on which they were inscribed were set up along the streets and squares, in the most frequented and most venerated spots of the sacred island, offering to all beholders the proofs of the wealth of Apollo and the piety of his worshippers.

Before the excavations of M. Homolle were made, a few fragments of inscriptions barely allowed us to catch a glimpse of the curious machinery of this administration of the sacred treasure. Now the records are abundant, and their variety infinite. There are general accounts and particular accounts, entries of expenses and receipts, leases and agreements for lending money; there are inscriptions which resemble the books of a business firm, in which are put down the incomings and the outgoings; there are complete statements of the position of the treasure, which form, as it were, the balance-sheet of the sacred treasury. Together with these documents which are really the books of a great banking house, there are the inventories which are not less curious; they alone give us an exact idea of what an ancient temple really was—that is, a sort of museum in which were accumulated objects of every kind for the cultus of the god and the adornment of the sanctuary. They show us also the internal arrangements of the temple, its sacred rites, and the things necessary for their celebration. Thanks to this mass of records, the administration of the sacred treasure is perfectly plain to us, and the museum which the temple once contained reveals to us all its secrets.

The treasure of Apollo, which, according to ancient

usage, was kept in earthen jars, was divided into two parts, and classed under two distinct heads: first, the sums delivered to the hieropoioi by their predecessors, or the funds in hand; secondly, the sums received during the year, or the annual income. The receipts forming this annual income were numerous, and of many different kinds. First came the landed property —the houses which the god let, the arable land and the pasturage which he farmed out; the rents and dues arising from these formed the first source of revenue. In the second place, there were the tolls granted to Apollo by the Delian State—a toll on the murex fishery in the waters of the sacred island, a toll on fish taken in the sea or in the sacred lake, a toll on the goods imported into Delos, a toll on the vessels anchored in the port. Lastly, as the income of the god more than covered his expenses, there was every year a surplus, which was put out at good interest in loans to towns or private persons. The reader will perhaps ask whether Apollo was an easy landlord and a complaisant money lender; the inscriptions allow us to satisfy ourselves fully on that point.

Apollo had houses to let suited to all tastes and all ranks: cellars and workshops, chemists' shops and baths, dwelling-houses large and small; he had also lands suited to all purposes, arable land as well as pasturage. But he was always anxious that the property, whatever it might be, should bring in a good profit. The managers of the temple were expressly recommended to see that all the property was always let, and on the best terms possible; and as the leases were drawn up according to a uniform model, the plan of which has been found, we can to-day ascertain precisely what those conditions were.

A lease was for ten years, and the tenant had to furnish good security. The rent was paid at intervals determined by the agreement, and any delay was punished by a fine; Apollo had no liking for unpunctual debtors. If the fine failed of its effect, severer measures were taken—the sale of the crops, the seizure of the animals and slaves; then, if the tenant were still a defaulter, there was a general seizure of his goods; lastly, his name was handed down to posterity inscribed on the stela where the names of those were placed who were debtors to the god. Apollo did not jest in money matters: if any condition of the agreement ceased to be fulfilled, the lease was at once cancelled by law; if the tenant left the island, or if he died, this gave the right to enter into a new agreement with another person; if he fell into poverty, the agreement was legally at an end. It must be admitted, however, that Apollo was not in fact very rich. His farms and rural domains bore picturesque and sonorous names, such as the Lakes, the Palm-trees, the Meadows; but they were few in number and not especially profitable. The rents of his houses brought in sixty-eight pounds yearly, and his farms two hundred and eighty pounds, which is but a poor revenue for a god. It may be added that as foreigners flocked to Delos and commerce increased, the farms became daily more difficult to let; farming, less remunerative, was less sought after and rents fell considerably. It is true that by way of compensation the rents of dwelling-houses rose to about the same extent; in spite, however, of their increased value, Apollo was not rich enough to allow his tenants to fall into arrears.

Fortunately the trade of money-lending served to swell the revenues of the god. The temple lent to towns as well as to private persons, and Apollo had a connection throughout the Cyclades among the borrowers of small sums. An agreement which has come down to us shows on what conditions the god consented to the loans made from his purse. He lent for five years, at 10 per cent.; the interest was paid annually, and repayment of the principal was demanded and enforced at the end of that time. As in such matters, and among Greeks, one cannot take too many precautions, the god took a mortgage on the property of the borrower, he demanded good security, and in addition, reserved to himself the right of seizing and selling any part of the debtor's property. The Delian State itself, when it borrowed from the sacred treasury, was also subject to Apollo's prudent caution. It was obliged to find substantial sureties for the repayment of the debt, and to give the temple a mortgage on the public revenue, and all these stipulations were duly recorded in an official document, which was deposited in the hands of a third person, generally one of the great bankers of Delos. Apollo, it will be seen, managed his fortune wisely, and took care to lose nothing which would bring him a profit.

In a well-managed house nothing is considered too small to be turned to account; and in the same way the god sold all the offerings in kind which were brought to his temple—geese, turtledoves, skins of the animals offered in sacrifice, even the dung of birds. Throughout the island, at the doors of the temples, boxes were placed to appeal to the generosity of the faithful; this was another small source of income for the god. Once a year these *treasuries* (as they were called at Delos) were opened, but it would seem that they did not bring in much.

Such were the annual revenues of the temple of the Delian Apollo. Collectors were appointed to receive them, and at the end of the financial year paid them into the sacred treasury. To these must be added the sum which the State paid to the temple towards the cost of festivals, choruses and dramatic representations; thus it will be seen that theatrical subventions are of ancient date.

When the hieropoioi had any payment to make, they took from the temple one of the earthen jars, and in the presence of the magistrates drew out the money needed. The expenses indeed were somewhat numerous. In the first place, the sacred buildings had to be kept up, and sometimes new buildings were erected; works of this kind formed the chief luxury of Apollo. In this too, of course, economy was carefully practised; contracts for the work were let by public tender. Several contracts relating to these buildings have been preserved; everything is minutely specified —the materials to be employed, the time allowed for the work, the guarantees to be given by the parties, the time and mode of payment—in short, the smallest details of the execution. To judge by these bargains, the contractor gained very little advantage over the god.

By the side of these charges, much the heaviest which

figured in the sacred budget, there appeared the cost of the temple worship. These were a multitude of small expenses which recurred from month to month. One day the temple was to be purified—so much must be paid for a pig. The altars must be crowned with flowers, and the sacrificial fire lighted—so much for wreaths, resin, faggots and charcoal. Festivals were to be celebrated in which choruses were needed—so much for lamps and other necessaries given to the performers. The maintenance of the buildings brought other expenses: sponges for washing, linen stuffs, nitre, wax, pitch for cleaning ; then there were incense, salt, vinegar, and especially oil, of which enormous quantities were used. Lastly, there were the office expenses for the purchase of paper and engraving of stelæ.

This was not all ; the salaries of those employed in the temple service must be paid. A flute-player, for example, received 120 drachmae yearly for her food, and 20 for her dress. Three musicians received 120 drachmae for food, and 16 for clothing. Each temple had neocori or sacristans, receiving 120 drachmae yearly ; the temple of Apollo alone had three neocori, but they were not so well paid. Then there were the architect, who received 120 drachmae, the secretary, who had 80, and others besides, all of whom were paid from the sacred treasury.

Nevertheless the budget of Apollo always showed a surplus. The annual income amounted to 27,000 drachmae, the expenses to about 21,000. There was therefore every year a surplus of 6000 drachmae, which was put out at good interest. In a balance sheet of the beginning of the second century the sum in hand is more than 60,000, the total treasure nearly 92,000 drachmae.

The wealth of the temple did not consist in money and land alone. There were also a multitude of offerings, which represented a considerable sum. We have already seen how popular the worship of Apollo was, above all in Eastern countries, and with what pious ardour kings and private persons brought their homage and their offerings to the altars of Delos. These were not generally casual gifts, given by chance and due to a liberality which would not recur; each year the return of the same festivals brought the return of the same embassies laden with splendid presents. They were, so to speak, perpetual pious foundations which the princes of the East had established at Delos, and on the solemn feast-days celebrated in their honour and called by their names, the Ptolemies, the Antigoni, the Demetrii, would have appeared ungracious indeed, if they had not regularly shown their gratitude by some magnificent present. Private persons and rich merchants who came to Delos emulated the kings, and lastly the Delian community itself made rich presents to the god to whom it owed everything. As long as the prosperity of the sacred island lasted the stream of offerings never slackened, and an epitome of the whole history of Delos is presented in the catalogues which enumerate them. To preserve these treasures and keep them in order was no easy matter, and to the difficulty it presented we owe the curious inventories which give us so much information as to the Delians.

The magistrates, who had the charge of the finances, were also entrusted with the care of the sacred offerings, and with the duty of keeping them intact and in good preservation. Hence arose a somewhat complicated system of book-keeping with regard to their reception. When one of the faithful brought a gift, the priest accepted it as the representative of the god, inscribed it in an entrance register, and assigned it a place in the temple. Generally all the offerings of the current year were placed in a group together, and it was only at the end of the year, at the time of making the general inventory, that they were placed with the rest of the collection.

It is this general inventory that the inscriptions give us. At the end of each year, all the temples of the island were solemnly visited, for the offerings were not always deposited in the sanctuary of Apollo. Some were in the Artemisium, some in the Temple of the Seven Statues, in the Porinos Oikos, and in other places besides, and all these buildings and storehouses were under the care of the hieropoioi. In each case they called over the list of the offerings according to the last inventory, and carefully verified it, lest anything might be missing. In order to prevent any error, each object was minutely described, and the place where it stood carefully stated ; for example, if it was near the door, or by the right or left wall, and besides all this any peculiarities in it were noted, and its weight given. The retiring hieropoioi were anxious to show that they were handing over the treasure exactly as they received it, the incoming officials wished to prove clearly in what state they took charge of it.

The many-sided interest of these inventories may easily be imagined. In the first place, they allow us to represent to ourselves the interior of an ancient temple.

The vestibule, like the shrine, was filled with offerings exposed to view. Some were hung on the wall, others stood on the ground or were placed on the sacred tables, but the greater part were ranged on shelves or stands. Fragile or valuable objects were kept in cases, small articles and coins in vases. Lastly, articles of the same kind, or the same material were usually arranged in one group; thus the different materials—gold, silver, bronze, iron, wood, marble—of which the offerings were made, formed separate classes.

It is easy to see that the constant increase in the number of offerings was a source of difficulty; some things were lost sight of, others, too closely pressed together, were injured. When this was the case a general rearrangement was made. Sometimes, in order to gain more space, a part of the offerings was sent to another temple, sometimes a part was melted down. In spite of all the precautions taken, the things, many of which were used in the ceremonies of the temple, wore out in the end, and they were not usually repaired. Thus accumulated little by little a heap of pious *débris*, and when these fragments became numerous enough to be in the way, they were got rid of by melting them down. In the same way, at the present time, in wealthy Greek churches, the numerous ex-votos are periodically melted down.

It is interesting to note the kinds of objects of which this treasure consisted, and this study makes us understand the manner in which an ancient temple was arranged and decorated. The separate articles were not grouped at random; on the contrary, an artistic feeling presided over their arrangement, for what was most desired was that the general effect should be beautiful and worthy of the god. With this end in view, the crowns were hung on the walls; similar in form and arranged in regular lines, they formed as it were a brilliant frieze all round the temple, and the costly stuffs and jewels served to adorn the statue of the god, while the vases were used to enhance the splendour of the solemn festivals.

Apollo possessed, in fact, a complete establishment. In the first place, he had his service of plate, composed of vessels and dishes of all shapes, generally in gold or silver, among which the most numerous were the phialae, which were dishes or cups with very wide mouths, and infinitely varied in decoration. The temples of Delos possessed large quantities of them; the Temple of Apollo alone contained more than sixteen hundred, plain or decorated, chased or embossed, adorned with fruit or foliage, animals or figures, gilded, damascened, or incrusted with precious stones. Then there were vessels of every kind, of every name, of every shape, of every size; drinking vessels and vessels for libations. There were coffers, censers, lamps, candelabra, lustres, stands for the offerings, couches on which to exhibit the statues, tripods, and I know not what besides; in short, a complete equipment. After the plate came the wardrobe, a mass of stuffs and jewels, of head-dresses and crowns, of gold-embroidered and purple tissues, which served to deck the image of the god, or to clothe his priests. The statue of Apollo wore a crown on its head, and the treasury also contained a second diadem for its use. A ring shone on his finger. The Graces were crowned in the same way, and another statue also had a complete wardrobe and set of jewels. There were also multitudes of crowns, rings, bracelets, necklaces, chains, pins, earrings, enriched with precious stones or adorned with engraved gems, brooches, scent boxes, rouge boxes, fans, ivory flappers encrusted with gold. All these things made their appearance at the solemn festivals, and their effect must have been marvellous. Besides all this there were instruments connected with different trades, sounding-leads, heralds' staves, bows and quivers, swords and helmets, anchors and rudders, all ex-votos consecrated by the faithful in memory of the dangers they had escaped. Lastly, there were ingots of metal, coins, and, above all, works of art, which formed one of the most interesting categories among the offerings.

It is true that this part of the catalogues by no means corresponds to the extraordinary riches found at Olympia; there the small bronzes are counted by thousands, and large statues are not rare. At Delos, on the contrary, hardly any objects of this kind have been discovered, and they are very rarely mentioned in the inventories. Nevertheless, a few curious details may be noticed: for example, the temple possessed several painted works, such as articles of furniture with painted panels and pictures, among which were some portraits. The treasure also included statuettes of men and animals, in gold, silver, or wood. Lastly, although life-sized statues do not in general appear in the catalogues, yet we can gather some interesting information with respect to them. In the temple of Isis was an Egyptian statue with a child on its knees; in the

Heraeum there were stone images wrapped in linen stuffs; elsewhere there were representations of Artemis, Aphrodite, and Serapis, and in the temple of Agathe Tyche, a marble statue of the goddess, holding in her left hand a gilded cornucopia, in her right a sceptre, and wearing on her head a crown set with precious stones. Lastly, the inventories mention a statue of Apollo, and we can, thanks to them, reconstruct the image of the god as it stood in his temple, as the old masters, Tectaeus and Angelion, carved it, holding a bow in the right hand and bearing the three Graces in the left. In the inventories of the second century B.C. allusion is made to this group, and to the golden crowns with which Queen Stratonice adorned the four figures; it is therefore probable that at that moment the primitive statue was still piously preserved in the temple, and it will be seen that it was of gold, either used in conjunction with ivory, or applied to a statue of wood.

V.

Nevertheless, these notices scattered throughout the inventories would give us but a faint idea of the artistic riches of Delos, if the excavations had not restored to us a certain number of statues. Certainly this archæological booty is very meagre in comparison with the rich series of inscriptions discovered in the exploration of the island, in comparison also with the multitude of works of art which once covered the shores of Delos. In ancient days the traveller landed in the midst of a crowd of statues, and even in the fifteenth century

Bondelmonte counted thousands of marble fragments scattered along the strand. To-day but very few monuments have escaped destruction, but notwithstanding their scanty numbers, they are interesting, and of the greatest importance for the history of archaic art.

The reader will remember the statues discovered on the Acropolis of Athens, and the manner in which they enable us to study the successive transformations of one and the same artistic type. A series of very similar works has been found at Delos, which is not less interesting, and still more precious, for the statues are of more ancient origin. In fact, by a singular good fortune, if Delos is comparatively poor in archaic inscriptions, she is rich in very ancient sculpture, belonging to the oldest artistic schools of Greece, those of Naxos and Chios.

As early as the middle of the seventh century B.C., these primitive schools were in a flourishing state. Even then they had carried the art of working in marble sufficiently far to show that they were no longer beginners; even then they had created two artistic types, which primitive Greek statuary adopted and popularised. One is that of the man, nude and standing motionless, the left leg slightly advanced as though on the point of stepping forward, his arms hanging, touching his sides, which we meet with at Thera, at Orchomenus, at Actium, and in other places besides, and which we shall study more closely when describing the excavations at the temple of Apollo Ptoïos. The other is the female type, specimens of which are found almost everywhere, at Eleusis, at Paros, on the Acropolis of

Athens, and which we have already had occasion to describe. Both are met with at Delos; it was in this guise, at the end of the seventh century, that the old Naxian masters represented Apollo and Artemis. The ancient statue of the god has not come down to us, but on the pedestal which bore it we can at least read the name of the sculptor who modelled it— Iphicartides of Naxos; this is the earliest artist's signature that we possess. However, to make up for this, the ancient image consecrated to the goddess, as the inscription cut in the stone states, by Nicandra of Naxos, has been preserved; it is the oldest work of Greek sculpture in marble which is known at present. In this primitive work, which dates from the seventh century, great delicacy is naturally not to be expected. It is impossible to imagine anything more naïvely uncouth. No shape is visible beneath the stiff covering in which the figure is swathed, the torso resembles a geometrical outline, rather than a human shape.* The face is quite flat, the different planes are scarcely distinguished from each other, everything in this rude and unskilful attempt shows the imitation of a wooden model, but for these very reasons it is interesting. It brings before our eyes the old *xoanon*, rough-hewn, or barely that, similar to the type of the earliest idols made in the likeness of a man, and it

Xoanon of Nicandra (Central Museum, of Athens).

* Homolle, "Statues archaïques de Délos" (*Bulletin de Correspondance Hellénique*, III., 101).

also shows us how the sculptor of Naxos, in those distant ages, represented the powerful goddess Artemis.

In opposition to the Naxian school, Chios offers another not less curious representation of the goddess.

Winged Victory of Delos (Central Museum, Athens).

It is the figure of a woman with a diadem on her head and wings on her shoulders and heels. Her attitude is strange; it is difficult to say whether she is in repose or walking, whether she is standing or kneeling; it is by this undecided and conventional attitude that the sculptors of the archaic school tried to express rapid motion. Her drapery is heavy and massive; the upper part of her body is enclosed in a sort of wooden cuirass or corset, but the head, already treated with more care, shows, in spite of the sculptor's want of skill, many of the distinctive qualities of Greek art. It is true that the imperfections are many, and the sculptor's inexperience is evident, but real technical skill is already shown—an attempt is already visible to give some delicacy to the modelling, and some expression to the face, on which there appears, as M. Homolle says, "the singular charm of that somewhat uncouth grace which distinguishes the early masters." On the base of the statue are engraved the names of the sculptors of the sixth century who modelled it—Micciades and Archermus of Chios.

Under these influences, which were transmitted from the East through the islands, sculpture developed at Delos as it did on the Acropolis, and a series of ten statues, which, by their attitudes and costume, exactly recall the Athenian marbles, enable us once more to follow the successive modifications of the type of the goddess Artemis, from its first appearance down to the latest days of archaic art. It is true that the Delian goddesses have not the incomparable brilliancy of the maidens of the Acropolis, the bright colours which adorned them have disappeared or faded, in most of them the head is wanting, but they are none the less interesting for the student of early Greek *technique*. It is unnecessary to repeat here the observations which have already been made with regard to the Attic marbles; between the statues of Delos and those of Athens the kinship is close, and the development of the two series proceeded on the same lines. At Delos also, the type was early fixed in its essential features, but each master in his turn improved it by modifications of detail, and thus the form became more elegant, the attitude more graceful, the draperies more flowing. Place side by side the primitive *xoanon* and the latest of these archaic figures; compare the first and the last steps of this long progress, and the immense distance traversed between the middle of the seventh and the beginning of the fifth centuries will be seen at a glance. It is true that even in these later statues archaism still reigns supreme, but remarkable technical skill is already visible, and a very delicate sense of form; they are works of art, and the art is sometimes almost noble.

The reader will, no doubt, ask whence came these statues, most of which have been found at one spot, in the enclosure sacred to Artemis. The inventories often mention the "Temple of the Seven Statues," which is considered to have been the primitive shrine of the goddess, and seven is exactly the number of the statues found in this part of the island; it is therefore probable that the excavations have restored to us the very works which formerly adorned one of the most ancient temples of Delos.

Apollo, as we have seen, had his statues at Delos as well as Artemis, but, less fortunate than his divine sister, he is not so well known to us as she. Of the statue of Iphicartides only the pedestal remains; of the colossal Apollo consecrated in the sixth century by the Naxians there only remains the base, with fragments of the body and thighs. Add to these an archaic head after the style of the Apollo of Thera, and two headless statues, which date from the fifth century, and represent the god in the customary attitude which is well known to us through the marbles of Thera and Orchomenus; with these exceptions nothing more has come down to us of all the numerous images erected to the Delian Apollo by his many worshippers.

It is not necessary to dwell long on the few other marbles, portraits, statues, or images of deities, found at Delos, or on the fairly numerous statues of the third and second centuries B.C. All the famous artists of the time were represented by some work in the sacred island; the *schola* of the Italians in particular was full of these monuments. One of the most remarkable is the statue of a warrior, the work of Agasias of Ephesus, a kinsman of the author of the Borghese "Gladiator," in the Louvre. Together with this master the inscriptions also name for the third century and the end of the second, Teletimus and Eutychides; a little later Polycles and Dionysius, Demostrates of Athens, and Aristandrus of Paros. The greater part of these artists were Athenians, and it is quite comprehensible that when the island had once more become the property of Athens, many Attic sculptors sought their fortune in this rich and luxurious town, where commissions must have been many and lucrative.

Without dwelling on these works of the second rank, we must mention another work of decorative sculpture found at Delos—the statues which adorned the pediments of the temple of Apollo. At the foot of the sanctuary, in front of the eastern and western façades, there have been discovered among other fragments six statues of the same size which evidently formed part of the decoration of one building; they are in fact works in the same style, if not by the same hand. These marbles were evidently placed against a wall as a background, for they are very carelessly worked behind, and traces of the iron fastenings are still to be seen which kept the figures in their places; they certainly served, therefore, to fill the tympanum of a pediment. It is not easy to determine what was the subject represented. On both façades of the temple was a scene of abduction: on the west front the principal group, which occupied the centre of the pediment, represented a woman carrying off a boy; on the east front the chief group consisted of a man carrying off a maiden in his arms. Around were placed statues of young girls calling for help, or flying before the ravisher. But if it is possible to reconstruct the scene almost in its entirety, we cannot explain it, and the legend of Apollo is silent with regard to the episodes which adorned the pediments of his temple.

We must therefore confine ourselves to pointing out the artistic value and the style of these marbles. They are of remarkable merit. The group of the man carrying away the girl is specially noticeable; the torso of the woman and her head, which is bent, are excellent. The group is full of motion and life: "In addition to exquisite beauty it possesses energy and power, and above all a certain boldness in its composition and in the attitude of the figures."[*] Two statues of young girls are not less worthy of attention. Injuries to the marble

[*] Homolle. *Bull. de Corr. Hell.*, III. 525, 526.

sometimes cause the figures to be depreciated, as it gives the appearance of defects in the workmanship; but as a matter of fact they are of great beauty, the modelling is very delicate, the bodies, full of life, unite strength and beauty, the drapery is treated with a light and skilful hand. It is no doubt true that they are not works of the first rank, nor of the best days of the Attic school; doubtless they possess "neither the serene majesty of the pediments of the Parthenon, nor the refined beauty of the balustrade of the temple of Wingless Victory"; such as they are, however, they cannot be placed later than the first half of the fourth century B.C., and they merit an altogether honourable rank among the works of sculpture which Athens produced at that epoch.

Such are the chief results of the excavations at Delos. In the history of the conquests of archæology its name should hold a place as considerable as Olympia and Pergamum; for these excavations do honour to the school at Athens which has directed them, and to France who has carried them out.

CHAPTER VI.

THE EXCAVATIONS AT THE TEMPLE OF APOLLO PTOÏOS (1884-88).

Holleaux, "Fouilles au Temple d'Apollon Ptoïos" (*Bull. Corr. Hell.*, 1885, '86, '87, '88, '89 and '90).
Collignon, "L'Apollon d'Orchomène" (*Bull. Corr. Hell.*, 1881).
„ "Torses Archaïques en Marbre provenant d'Actium" (*Gaz. Archéol.*, 1886).
Rayet et Thomas, *Milet et le Golfe Latmique*. Paris, 1878.
Brunn, *Arch. Zeitung*, 1876.
Fränkel, *Arch. Zeitung*, 1879.
Furtwaengler, *Arch. Zeitung*, 1882.

To the east of the wide low plain which a few years ago was still covered by the marshy waters of Lake Copaïs, and at a short distance from the village of Karditza, the ancient Acraephiae, lie the recently-discovered ruins of the temple of Apollo Ptoïos. There, on the side of Mount Ptoïon, on a lofty terrace supporting the building, rose the Doric temple consecrated to the god, whither an oracle famous throughout all Boeotia continually attracted a crowd of worshippers. The foundation of the temple was attributed by legend to Apollo himself, who, according to Pindar,[*] as he was wandering over the earth chanced to stop one day on the summit of Mount Ptoïon. "Then, gazing over all the plain that lay beneath him, he cast down the

[*] Pindar, *Fragments*, No. 70 (Bergk., 3rd ed., pp. 78, 79; 4th, 101 and 102).

mountain mighty stones to form the foundations of his temple."

The reputation of the oracle seems to have stood very high in the early days of Greece, for numerous fragments of ex-votos in bronze, tripods, vases, vessels of every kind, and especially precious statuettes in the archaic style, pious offerings consecrated by votive inscriptions to Apollo, have been found in the course of the excavations. The fame of the sanctuary spread so widely that pilgrims came from distant Asia to consult the oracle, and the connection between its priests and the countries on the other side of the Aegean was so close that they sometimes gave their responses in Carian. Consequently, when the Persian invaders descended upon Greece, Apollo, for whom moreover the Achaemenid princes seem to have felt an especial regard, weathered the storm; nay, more, Mardonius did not disdain piously to seek for guidance from the Ptoïan god. In later times, after the cruel punishment inflicted by Alexander upon Thebes, the temple was less fortunate; popular favour seems to have deserted it, and it was not until the second century that it regained a lasting splendour. Then the sacred games, held every four years, once more brought worshippers in crowds to the sanctuary, and the neighbouring deities warmly commended the new institution, another instance of that mutual support of which we have already had examples. The Amphictyonic council proclaimed a sacred truce similar to that which had formerly made the greatness of Olympia, declared the inviolability of all who should come to take part in these games, and allowed the right of asylum to the sanctuary of Apollo. The celebrated oracle of Trophonius at Lebadaea was not less courteous towards the god of Mount Ptoïon, and decreed in an oracular response, as solemn as it was obscure, that the new festival should be recognised among the sacred games. Accordingly the shrine appears to have enjoyed uninterrupted prosperity for some centuries. Most of the Boeotian towns esteemed it an honour to take part in the celebration of its games and to set up votive offerings in its temenos; and when, every fourth year, the town of Acraephiae sent ambassadors at the command of the prophet of Apollo to invite its neighbours to the festival, they came from all parts to witness the contests in music, singing and poetry which took place in the theatre, to join in the solemn dances and processions, and to share the magnificent banquets offered to the assembled throng in the course of the games. Nor did the Athenians themselves, or the dwellers in the towns of the Peloponnesus, such as Argos and Mantinea, or even the distant cities of Asia, disdain sometimes to take part in the festival. Wealthy private individuals enhanced the splendour of the games by their gifts. The Boeotian league, which in conjunction with the town of Acraephiae exercised authority and control over the temple, paid solemn homage to the god, and the faithful crowded to consult the prophet of Apollo.

In the Roman period the festival seems still to have been celebrated with especial splendour, and among the agonothetae who at this time presided over the games, we find the wealthy Epaminondas of Acraephiae, one of the benefactors of the town and temple. It was he who caused to be engraved and set up in the temenos of the Ptoïan temple the strange speech made by Nero at Corinth when he restored their liberty to the Greeks, a remarkable relic of imperial eloquence which we have recently had the good fortune to recover.

It was towards the close of the year 1884 that M. Maurice Holleaux, a member of the French school at Athens, began to excavate the temple of Apollo Ptoïos at Perdicovrysi, above Karditza, and his successful explorations, carried on through several successive seasons, soon yielded the most brilliant results. Not only do we owe to these investigations numerous inscriptions which have thrown a flood of light on the worship of Apollo, the administration of his temple, and the games celebrated in his honour; but M. Holleaux has discovered among the ruins of the sanctuary, which have been so far identified as to allow of an accurate restoration of the building, a rich and important series of statues in bronze and marble which throw new light on the history of the early development of archaic art in Greece.

The importance in connection with the history of art possessed by those remains which enable us to follow the successive transformations of the same type, is well known, and the excavations at Delos and at Eleusis, as well as the discoveries on the Acropolis of Athens, have shown by what prolonged and ingenious efforts, by what delicate and instructive variations, the early Greek sculptors gradually perfected that uniform type which for many years served to represent the female form.

By the side of these female figures, whether they

represent divinities or simple mortals, the old Hellenic sculptors created a male type also, which in the workshops of the archaic period long served for those beings of the male sex, whether gods or men, whose forms the artist wished to represent. Numerous examples of

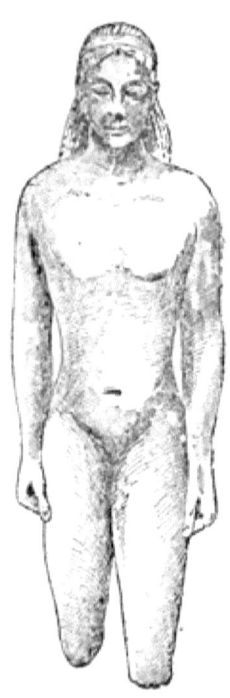

Apollo Ptoïos (Central Museum, Athens).

this type found of late years at Orchomenus and at Thera, at Delos and at Naxos, at Actium and at Tenea, and at other places also, had shown us before M. Holleaux undertook his excavations, that it was common in the archaic period to the greater number of schools, and had enabled us to ascertain its characteristic features. A beardless man, entirely nude, stands motionless and stiff, his arms hang down, one leg is slightly advanced, his head is rigid, and his long hair, thrown back, falls on his shoulders, and is there spread out in the shape of a fan. Nevertheless, in spite of the importance of the examples which had been preserved, this series of archaic statues, in which it was usual to recognise representations of Apollo, was still too incomplete to allow of a thorough investigation of the type. The excavations at Perdicovrysi, by enriching this interesting collection with some valuable figures, gave new data for the discussion of the question, and introduced into this difficult investigation a precision for which we had not dared to hope.

They have presented us, in fact, with at least eleven copies of the same model, mutilated, it is true, but none the less singularly instructive; and these figures, which extend from the end of the seventh century to possibly the third quarter of the sixth in almost unbroken succession, show the steady but constant progress by which the early masters perfected the representation of the male type. The invention of this type in plastic art is generally attributed to the Cretan sculptors who were at work in the Peloponnesus towards the end of the seventh century. It was indeed to be expected that the first efforts to represent Apollo should be made among the Dorians, who honoured him as their national divinity. Under the influence of these old masters, Dipoenus and Skyllis, the primitive type, which was doubtless carved in wood, spread throughout central and northern Greece, and all the archaic statues which have come down to us are reproductions of it. Nevertheless for a long time two principal groups have been distinguished among these very early sculptures: the Apollo of Thera and that of Tenea, by the general structure of the body, the bold relief of the face, and the pinched, affected smile which twists their lips into a grimace, differ unmistakably from the statues of Orchomenus and Actium. Both groups may undoubtedly be referred to the same primitive type, but in the different countries into which it was introduced it was freely modified in accordance with local taste, and became the starting-point of independent artistic development. The statues from Perdicovrysi prove this in a remarkable way; the analogies they offer with the type of Orchomenus are so many and so characteristic that it is impossible not to recognise in them the unmistakable stamp of an indigenous art ; and thus these valuable marbles, which have already shown us with remarkable clearness the method by which Greek sculptors slowly perfected one of the favourite types of primitive studios, render us a further service by making us acquainted with a unique series of relics of that early Boeotian art to which we had hitherto been strangers. This plastic type was no doubt remotely derived from wooden models. These rude attempts inspired the sculptor of that fragment in calcareous tufa which is one of the oldest figures found at Perdicovrysi, and is a work of the most rudimentary kind, recalling

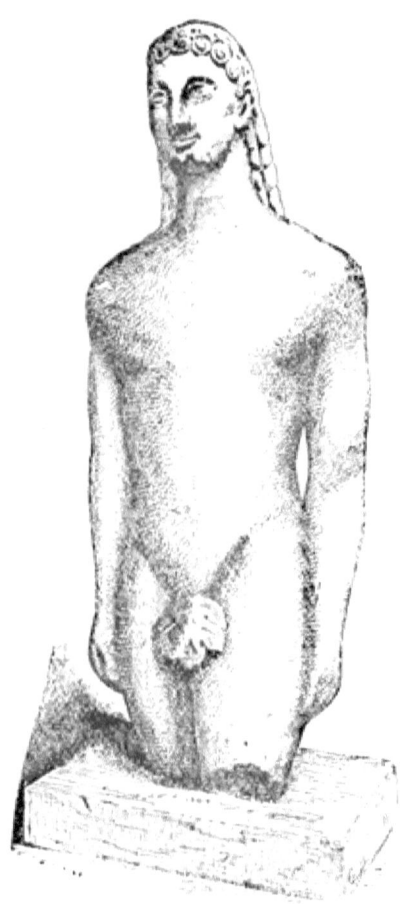

Apollo of Thera, front view.

the shapeless idols, roughly hewn into the form of a pillar or a column, which were objects of adoration to the simple piety of early Greece. The technique of wood-carving is still evident in the Apollo of Orchomenus, and even though marble begins to replace the original material, though the shapeless and sexless piece of wood begins to assume a vague likeness to the human form, we still trace, in this very early statue, the remembrance, not yet effaced, of the idol carved from a narrow plank. The old master indicates the modelling of the torso by an almost geometrical precision in marking off his planes and by broad, smooth surfaces cutting one another sharply, and thus unconsciously applies to marble the rudimentary technique of wood-carving. The same inexperience is shown in the flat and heavy profile, while the disregard of anatomy proves that the sculptor was guided by the tradition of a school rather than by direct study of the human frame; nevertheless, in spite of the clumsy execution, a naïve effort was made to transform the primitive *xoanon* into a living figure, and in the framework of the body we can already mark the relative correctness of the proportions. The Bœotian sculptors must for a long time have made use of the type thus perfected, and several fragments found in the temple of Apollo Ptoïos are evidently

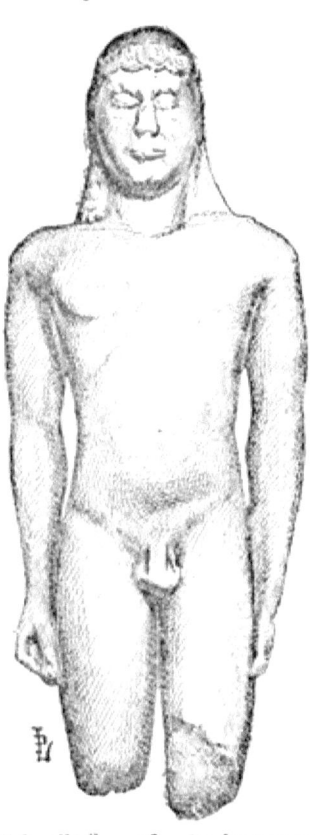

"Apollo" of Orchomenus
(Central Museum, Athens).

contemporary with the Apollo of Orchomenus. Still this early art was not stationary. Every master endeavoured by some ingenious variation to give more pliancy and animation to the old type, and the Boeotian sculptors went from one improvement to another by insensible gradations, but with a rapid step. Take, for example, the curious stone head found at Perdicovrysi in 1885, which shows by its bold execution its sharp ridges and its hard modelling, that it is closely related to the Apollo of Orchomenus, and that the memory of the technique of wood-carving was still fresh. It is full of faults, it is true, but it is also full of effort and of promise; in spite of the artist's inexperience we feel his ardent desire to observe and imitate nature, and a certain conscientious sincerity to which the sculptor of Orchomenus was an utter stranger. Thus, by a succession of imperceptible improvements, which the discoveries of M. Holleaux enable us to follow in, the most minute detail, the primitive type was softened and transformed. Look at the long series of ancient statues which decorated the temple of Mt. Ptoïon. All, with a few variations, reproduce the same type, and yet each of them marks a stage in the slow transformation. Here is one of the best preserved among them, which closely recalls the statue of Orchomenus. The attitude is the same, the structure and the face are similar, as also the technique; and yet it makes a real advance. It is true that the figure is still cold and rigid, the anatomy is very superficial, and the design conventional; but the artist already feels a wish to please, he has taken some thought for grace and elegance, and has felt some aspirations which his unpolished predecessor never knew; and if his work is wanting in life, at least it is not without charm in its uncouth simplicity.

Let us wait a few years, and we shall find that this persevering effort to attain plastic truth is almost crowned with success. In a fine marble torso, which evidently belongs to the second half of the sixth century, the contours are already fuller and more pliant, the flesh has more substance; but above all—and this is the great innovation—the figure has thrown off its traditional immobility, and the arms, which have hitherto adhered closely to the sides, are now detached from them and extended in front. There is still much inexperience visible in the figure, and the anatomy remains very superficial; but in this statue, as well as in the Apollo of Tenea, which dates from the same period, we feel that worn-out traditions are giving place to a better comprehension of the truth.

"The works of this period," says M. Holleaux, "please us more by what they allow us to divine than by what they actually show us; they

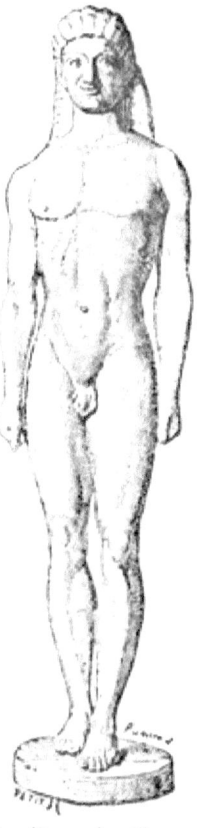

Apollo of Tenea (Glyptothek, Munich).

derive a remarkable attraction from the promise they contain, already visible under the imperfection of the present, from the future they foretell, and from the speedy and confident expectation they excite of the finished works which are to follow."

The most striking quality in these works of the Boeotian school is the honesty of their workmanship, which without conscious preference, without systematic design, without any trace of the mannerism so plainly marked in the Apollo of Tenea, pursues the search after plastic truth rather by imitation of nature than by systematic study of anatomy. Very marked also is the desire to vary the attitudes of the figures and to render their movements more supple, and the tendency to break the tedious monotony of traditional types by innovations which, it must be granted, were sometimes uncouth.

No doubt the master's hand often fails to express his thought—his skill is not equal to carrying out the ingenious boldness of his conception; and yet these works are charming in their naturalness, in their frank spontaneity, and in the effort, sometimes successful, of which they show the trace. It is these qualities which constitute the interest of a remarkable marble head, a work of the second half of the sixth century, which in its ripened archaism has a penetrating charm.

We can distinguish in it all the qualities peculiar to the Boeotian school; the oval shape of the face, the somewhat prominent, strongly-marked profile, the very prominent eyes, and especially the discreet smile upon the lips, all show the artist's sedulous desire to depart from the conventional type. In addition to this, the workmanship is often excellent, the modelling soft as well as exact, and the effort made to give expression to the face has not been unsuccessful. The whole face gives an impression of gentle serenity, and almost of majesty; it deserves a distinguished place among the archaic sculptures of Boeotia.

By the side of the motionless figure of Apollo, which is represented in the long series of statues from Perdicovrysi, the old Greek sculptors, in the eagerness of their desire to break the monotony of sacred tradition, conceived another type of the same deity, beardless and nude like the preceding, but freer and more lifelike in its attitude. In this new representation both arms below the elbow are detached from the body and extended, and each hand holds an attribute, generally a bow and arrows. Long before the day when a famous statue, the Apollo sculptured by Canachus for the temple of Miletus, gave unheard-of popularity to this design, the masters of the archaic school took delight in reproducing it, and a very ancient bronze discovered in the Ptoïan temple, rude in workmanship and almost childish in its clumsiness, with heavy, flattened contours and a grotesque and almost shapeless face, affords one of the oldest examples of this second class. Several other bronze statuettes belong to the same series; but while the large marble statues set up in the temple precincts are almost all the work of native sculptors, the small bronze figures, much more easily transported, were brought from many different places to adorn the temple of the god. They belong to different artistic schools, according to the dedicator's nationality, and in this way they are singularly instructive for us. One, which is still very archaic in appearance, but strong and skilful in its modelling, exact to minuteness in its workmanship, delicate and refined, but somewhat frigid and lacking in freedom and naturalness, recalls by striking analogies in attitude, type and coiffure, a statuette of Apollo discovered at

Naxos and now in the Berlin museum; both works belong to the latter part of the sixth century, and the dryness of their manner, their somewhat measured exactness, would seem to show that they came from the Peloponnesus and are the work of Dorian hands. Another, on the contrary, although it reproduces the same design and dates from almost the same period, is distinguished from its neighbour by radical differences of workmanship. Its contours are square and massive; the very strongly-marked type of face recalls in a remarkable way the marble faces of the Boeotian school; and, in spite of a *naïveté* which provokes a smile, in spite of evident lack of power and skill, the work charms by its naturalness and freedom from mannerism. We find in it all the characteristics of primitive Boeotian sculpture: the striving after originality, the freedom from convention, the remarkable sincerity of the work; and in these respects the statue forms a striking contrast to Peloponnesian works. The latter undoubtedly display more science, and we sometimes find in them, as for example in the small female figure found in the temple of Apollo Ptoïos, which belongs to the same school, a curious mixture of intentional archaism and of almost finished elegance; but they are wanting in ease, their elegance savours of painstaking and attentive study, together with a timid respect for the conventions of their school. The Boeotian figure knows nothing of these delicacies; but, on the other hand, it possesses a native vigour and energy; it is a hasty and spontaneous work, crude, no doubt, but at least full of character and freshness. By these qualities it completes the conception which the marbles from the shrine of Apollo Ptoïos had already given us of the early sculptors of Boeotia. "They have," says M. Holleaux, "much less talent than feeling, little skill, but a great deal of sincerity. If their works are very far removed from plastic perfection, they have kept close to nature; they have little of the knowledge which is to be gained by study, but they are endowed by instinct with some of those qualities which cannot be acquired." *

Many of these statuettes are rendered still more interesting by the fact that they undoubtedly reproduce some work of a great master, possibly the Philesian Apollo, sculptured by Canachus of Sicyon towards the end of the sixth century for the Didymaeum of Miletus. Many ancient copies on a reduced scale of this celebrated statue, such as the Payne-Knight bronze and the Strangford Apollo in the British Museum, and especially the precious bronze Apollo of Piombino in the Louvre, give us a fairly exact idea of the work of the Peloponnesian sculptors. The god stood erect, beardless, and entirely nude, the left foot slightly advanced, the

The Piombino Apollo (Louvre Museum).

* Holleaux, *Bull. Corr. Hell.*, xi. 360.

arms raised from the elbow, the hands holding a bow and a fawn in the characteristic attitude created by the archaic sculptors ; the hair short and curling over the forehead, fell over the back in a thick mass, and hung in symmetrical locks upon the breast, the eyes were of silver, or of some precious stone inserted in the bronze. The sculptor of Sicyon was still hampered by the conventions of primitive art, and many traces of earlier harshness or stiffness are still visible in his work, yet in his Apollo he has refined the features characteristic of the class ; the sturdy figure displays a more careful study of nature, and if the expression is still somewhat commonplace and empty, as we also find it in the fine archaic head of Apollo discovered at Herculaneum, yet the progress is undeniably immense which in a few years has enabled these early masters to traverse the interval which separates the rude and coarse figure of the Apollo of Orchomenus from the work of Canachus.

The excavations on Mount Ptoïon have restored to us a statue in Parian marble, in which we cannot fail to recognise a copy of the bronze of Canachus. It is closely akin to the Apollo of Piombino and to the Strangford marble, but it is earlier in date than these two replicas of the Didymaean Apollo, and belongs to the period of transition which lies between the early archaic sculpture and that of the Aeginetan school ; its date may be fixed with certainty towards the end of the sixth century, somewhat earlier than that of the figures on the western pediment of Aegina, which it resembles in more than one point. We do not find in this statue either the freedom and breadth of style which mark the Boeotian school, or the grace and delicacy and intensity of life which distinguish Attic works. The value of the figure consists in its anatomical precision, in the careful and minute study of details, in the modelling of the muscles, in the sobriety of the workmanship and in the harmonious and simple construction of the whole; and in these points it recalls the style of the Aeginetan school. As in the Aeginetan marbles, the head is much inferior to the body; the work is less finished, the expression almost meaningless, and this also shows the manifest influence exerted by the old Peloponnesian schools. In this Dorian work, visibly inspired by the bronze of Canachus, it is interesting to find, not only a replica of a celebrated statue, but also one of the latest transformations which the type of Apollo underwent during the development of archaic Greek art.

In the complete work which M. Holleaux is about to issue on the subject of the excavations undertaken at the temple of Apollo Ptoïos, he will describe many other remains, which we cannot even name in the course of this study. All that we have tried to do has been to point out the great interest of these recent excavations, which throw such vivid light on the methods of archaic Greek art, and which reveal in an important series of remains, the slow transformation through which one of the types dear to the early sculptors has passed, as well as the patient efforts of one of those ancient schools of the Peloponnesus and Boeotia, whose diligent study and original qualities paved the way for the full development of Hellenic art.

CHAPTER VII.

THE EXCAVATIONS AT OLYMPIA (1875—1881).

Books of Reference:
>   Curtius, Adler, Treu, Doerpfeld, "Ausgrabungen zu Olympia."
>   "Funde von Olympia."
>   "Inschriften aus Olympia" (*Arch. Zeit.*, 1876—1881.)
>   Boetticher, "Olympia," Berlin, 1886.
>   Flasch, Article "Olympia," in *Baumeister's Denkmaeler*, Munich, 1888.
>   Laloux et Monceaux, "La Restauration d'Olympie," Paris, 1889.
>   Prof. Gardner, "Olympia" (New Chapters in Greek History).
>   Brunn, "Die Sculpturen von Olympia," Munich, 1877; "Paionios und die Nordgriechische Kunst.," Munich, 1876.
>   Furtwaengler, "Bronzefunde aus Olympia."
>   Curtius, "Altäre von Olympia."
>   Reinach, "L'Hermès de Praxitèle (*Rev. Arch.*, 1888, t. 1).
>   For pediments, Six, in *Journ. Hell. Studies*, and literature quoted there.

I.

IN the district which the ancients knew by the name of Elis, on the western side of the Peloponnesus, there lies on the banks of the Alpheus a plain nearly 1100 yards in width, and surrounded on all sides by a line of low hills. The river, which has just descended through a narrow gorge from the lofty mountains of Arcadia, here slackens its swift, impetuous course, and disports itself in fantastic curves and windings, forming a number of small islands, and gathering volume from the tributaries it receives from the northern side of the valley, of which the most important is the Cladeus. It was at the confluence of these two rivers, at the foot of the wooded slopes of Mount Cronion, or Olympus of Elis, that there rose in ancient times the famous temple of Olympia; and it was in this little plain, round the sanctuary of Zeus, that the most famous games of ancient Greece took place. If we are to believe the naïve traditions handed down by the priests, the origin of these games was to be found in the distant days of the Golden Age, when Zeus was still an infant, and his father Saturn the monarch of the world. The gods themselves had not disdained to be the first to take part in them; Zeus had there disputed the sceptre with Saturn, and Apollo had vanquished Hermes in running and Ares in boxing. In later years, so ran the legend, Pelops, son of Tantalus, had given the Olympian games new splendour, after the famous adventure in which, favoured by Zeus, he had overcome King Oenomaus and wedded the fair Hippodamia. The story is well known; it was one of the most celebrated traditions of Olympia, and was immortalised by art on one of the pediments of the temple. Oenomaus, king of Elis, had a daughter Hippodamia, and the Delphic oracle had foretold that he would die on the very day of her marriage. Consequently the king was firmly resolved to reject all suitors for her hand, and devised an easy way of disposing of them: he challenged them to a chariot race from the altar of Zeus at Olympia to that of Neptune at Corinth. The race was long, many accidents might happen by the way, and as a rule the king's good spear rid him of his adversary in the course of it. When Pelops, son of Tantalus, presented himself in his turn, he was more fortunate, or rather, more cunning. His riches enabled him to win over Oenomaus' charioteer, who overturned his master, upon which the king killed himself in despair. Pelops then married Hippodamia and succeeded to the vacant throne, and in order to testify his gratitude to the gods, celebrated the Olympic games afresh with unaccustomed splendour. According to another legend, Hercules, after he had overcome Augeas, king of Elis, offered solemn sacrifices on the plains of the Alpheus in honour of his father Zeus; then with the ashes of the victims he raised an altar to the god, and celebrated the Olympic games for the first time within the sacred precincts, which he surrounded with a wall. Finally, in the ninth century, Iphitus, a king of the country, instituted the sacred truce, in order that the festival might take place in peace. Henceforward, by a solemn convention accepted by all the Peloponnesian states, the territory of Olympia became neutral ground, access to which was forbidden to any force of armed men during the festival, and the inviolability of which was placed under the joint guarantee of all the Hellenes. From this time the territory of Olympia was sacred, and its greatness assured; the obscure local festival of a district in the Peloponnesus became a national institution of Hellas, and the Eleans showed both gratitude and good sense when, in the entrance to the Temple of Zeus, the most conspicuous place in the sanctuary, they set up a statue of King Iphitus crowned by Ekecheiria, the goddess personifying the sacred truce to which Olympia owed its prosperity.

" Every one knows how the Olympic games became the national festival of Greece, and how their return at regular intervals of four years served as a basis for the only system of chronology common to all Hellas, that of the Olympiads. To a greater extent, indeed, than any of the other Greek games—more than the Isthmian or the Nemean, more even than the festival of Apollo at Delphi—the solemnities of the Olympian Zeus served as a national centre to the Greek world. Pilgrims flocked to it from every part: Dorians and Ionians, men of Athens, Sparta, and Thebes, all, in spite of their mutual rivalries and hatred, and in the midst of their most embittered struggles, forgot for a moment their ancient feuds when the festival came round, and lived together in harmony for a few days at this peaceful rendezvous where all could meet. The prizes won at the Olympic games surpassed in glory all other honours; and the mightiest men of their time, the kings of distant Cyrene, the tyrants of Sicily, the heads of the powerful aristocracies of Corinth, Argos, and Thessaly, the richest citizens of democratic states, had no higher ambition than to carry off the prize in the chariot-race at Olympia." *

Neither the difficulties nor the length of the journey, nor the imminent danger of foreign invasion, diminished the keenness of the desire to hold these "great assizes" of Greece; and at the very time when Leonidas was meeting a glorious death on the field of Thermopylae, his countrymen assembled at Olympia were holding undisturbed the solemn games in honour of their gods. The reason for this was that these games in the stadium, these victories in the arena, were of far different import among the Greeks of classical times

* Rayet, *Études d'Archéologie et d'Art*, pp. 43-4.

than among ourselves. The long training of the body which the Olympian victors underwent, the ardent devotion to gymnastics which fashioned these athletes and runners, were more than a mere amusement; they were an essential part of any thorough education, and almost a patriotic duty, since these exercises trained for their country a picked body of good soldiers and brave citizens. "With such men," it was said of the Olympic victors, "there is no need of walls"; and indeed the athletes who won the crown of wild olive were given equal honours with the most famous generals; they made triumphal entries into their native towns, and their victories seemed, as much as the greatest military successes, to be a visible sign of the protection of the gods.

We can therefore imagine the rapid development the Olympic festival underwent, and it must be added that to these great assemblies of Greece, to this international meeting-place of Hellenic piety, none came with empty hands. All were eager to lay their pious offerings at the foot of the altars of the Olympian Zeus; each city set up statues in honour of its successful athletes around the temple of the god, and consecrated to him, after a victory, as a tithe of the spoil, some monument whose pompous inscription might recall to future generations the glory and the piety of its founders. All the great political events which stirred the Hellenic world, every caprice of fortune, enriched the temple of Olympia with new offerings; kings, tyrants, towns, vied with one another in offering splendid gifts; the greatest artists of Greece spent their skill on these works, thousands of which covered the central space and the terraces of the sacred enclosure. This collection of votive offerings, this multitude of figures in bronze and marble, gathered together under the great trees which surrounded the sanctuary, formed an unrivalled museum, the finest and most famous in Greece. Even in the Roman period, although the festival had by that time lost much of its ancient splendour, pilgrims returned from Olympia dazzled by so many marvels, and by the multitude of statues—more than three thousand in number—which formed the silent and magnificent train of the Olympian Zeus. Pausanias, who in his ten books has described the whole of Greece and has devoted but one to the monuments of Attica, pauses in surprise and admiration before the treasures of Olympia, and devotes two whole books, perhaps the most interesting of all, to the enumeration of the works of art, the statues of divinities, heroes and athletes, the pious offerings, the altars, monuments and temples, crowded into its sacred precincts.

The renown of the Olympian games was so great that it even survived the triumph of Christianity. They existed, although shorn of much of their ancient glory, until the end of the fourth century—till the day on which the Spaniard Theodosius ascended the throne of the Caesars. This ardent adherent of Christianity waged merciless war against the fallen divinities of paganism, and in the year 393 A.D. the games were held for the last time. In the following year an imperial edict prohibited them for ever, and silence fell upon the fair and peaceful valley of Olympia, and around the temple of Zeus. The buildings were stripped of their more precious treasures, which were removed to Constantinople; among others the Zeus of Phidias, which had once tempted the dilettantism of Caligula, became the prey of the imperial rapacity, and was carried off to Byzantium, whence some years later it disappeared for ever in a great fire. Still the temples remained standing, as well as many of the monuments; it is even possible that the heathen priests were still rendering their accustomed homage to their gods in the deserted Altis when the Goths, under Alaric, invaded the Peloponnesus. Hemmed in by the imperial armies in the immediate neighbourhood of Olympia, in all probability the barbarians did not spare the treasure of the sanctuary. Notwithstanding this, however, in 426 A.D. the temples were still standing and intact, for in this year the Emperor Theodosius II. ordered the buildings of Olympia to be given to the flames.

The gods were now for ever fallen, but human efforts would have been unavailing to cast the gigantic blocks and enormous columns of the pagan buildings to the ground. Nature undertook to complete the task imposed by the imperial edicts, and the earthquakes, so frequent in the Peloponnesus, overthrew the buildings on which man had made no impression. In the year 522, and again on July 9th, 551, a terrible catastrophe spread destruction throughout the East: Patras, Naupactus and Corinth became mere heaps of ruins, and Olympia perished in the same disaster. During several years a wretched village continued to exist in the midst of the ruins, under the shelter of the Byzantine citadel which was built round the temple of Zeus after the invasion of the Goths; a church, a few poor houses, a few tombs recall the memory of these obscure inhabitants of the fallen Olympia, whose coins show that they remained there until the seventh century. Then follow silence and oblivion. The Slavonic invaders who took possession of the Peloponnesus, the French barons established in the peninsula, the Turks and the Venetians, passed by the tomb in which the sacred city of Olympia lay buried without a glance, and Nature, which knows so well how to heal the wounds made by time and violence, slowly cast its covering of sand and verdure over the ruins ; they had to be disinterred from a depth of fifteen or twenty feet beneath the soil.

The entire disappearance of the ruins of Olympia was long attributed to a great flood caused by the overflowing of the Alpheus. This froward stream has had a very bad reputation from ancient times, and the Greek legend, delighting to transform the great phenomena of nature into mythological tradition, told with what passionate impetuosity the river-god, enamoured of the nymph Arethusa, pursued her without pausing over the waters of the Ionian Sea to the distant shores of Sicily. The disappearance of Olympia was readily ascribed to the violence of this stream, and it was long hoped that as Pompeii was given back to us entire from beneath the ashes of Vesuvius, so we should find Olympia as it was in the days of its splendour preserved uninjured beneath the river-ooze. Alas ! our expectations have not been fulfilled. The Alpheus had but a small share in the destruction of these monuments ; it was the inhabitants of the Christian village established among the ruins who ruthlessly carried confusion into the buildings within the sacred precincts, while the land-slips, common in this part of Greece, have gradually completed the work of destruction. From the slopes of Mount Cronion, which overhangs Olympia on the north, enormous masses of earth have fallen on the buildings near it, the Cladeus has broken through the dykes by which the ancients regulated its stream, and has made its way across the Altis, lastly, the Alpheus has carried away the low-lying ground next the river; and thus, until the excavations, of the present century, were undertaken nothing was left of Olympia but a desolate and monotonous plain which presented no trace of its former splendour.

It was in the year 1723 that the name and memory of Olympia emerged for the first time from the darkness, and it was a Frenchman who first commended to public attention the field on which German science has since made such splendid discoveries.

In a letter dated June 14th, 1723, the learned Benedictine Montfaucon congratulates Cardinal Quirini, who had just been appointed Bishop of Corfu, and enlarges upon the interesting archæological discoveries which his appointment will enable him to make in Corfu, Cephalonia and Zante, adding:

"But what is all this in comparison with the discoveries which might be made on the opposite coasts of the Peloponnesus? There is the Elis of antiquity, where the Olympic games were celebrated, where numberless monuments were set up to do honour to the victors—statues, reliefs, inscriptions. The earth must be full of them; and what is peculiar to this spot is that, as far as I know, no one has yet made any explorations there."

Winckelmann afterwards expressed the same wish :
"I feel assured," he said, "that there is a harvest to be reaped in Elis which will surpass every hope, and that a thorough exploration of that district will throw a flood of light upon the history of art."

Not satisfied with uttering these prophetic words, the indefatigable archæologist, who as one of his contemporaries wittily said became a Catholic in order the better to study Rome, and would gladly have turned Mussulman in order to explore Olympia, preached a veritable crusade in favour of his belief, interested the princes of the Roman Church in the enterprise, and until his death in 1768 never ceased to converse with his numerous friends on his favourite plan. "What one earnestly desires," he said, "always becomes possible, and I have this matter as much at heart as my history of art."

Winckelmann's project died with him. But just then, for the first time, an Oxford theologian, Richard Chandler, visited the ruins of Olympia, and acquainted the scientific world with the fact that the remains of a large Doric temple existed in this remote corner of the Peloponnesus. After him the French traveller Fauvel, entrusted with an archæological mission by the Marquis de Choiseul Gouffier, identified these ruins for the first time with the temple of the Olympian Zeus. At this time, moreover, the sanctuary had become a quarry for the neighbouring village, whence, according to a deplorable custom which has been the destruction of many ancient buildings, they sought for building-stones, well-hewn and ready to hand. The signal had now been given; other travellers, who had wandered over the Peloponnesus between the years 1801 and 1808, proclaimed in their turn that the most splendid discoveries might be made there, while about the same time Quatremère de Quincy was at work on his book on the Olympian Zeus of Phidias, and by his advice, with the support of the Institute of France, Lord John Spencer Stanhope published in 1824 the first complete map of the district of Olympia.

Up to this time no excavations had been undertaken. It is again France to whom the honour belongs of undertaking the first systematic exploration of these ruins. It is well known that in 1829, during the Greek War of Independence, and immediately after the battle of Navarino, the government of Charles X. landed a body of troops in the Peloponnesus commanded by Marshal Maison. "The expedition to the Morea was accompanied by a scientific mission, as that of Bonaparte to Egypt had formerly been; and while the officers of engineers made a survey of the peninsula, and the naturalists studied its flora and fauna, a commission composed of archæologists and artists, at the head of which was the architect Abel Blouet, explored the ancient ruins."*

Among the points to which the Institute had drawn the attention of the mission, the temple of Olympia held an important place, and in fact the excavations undertaken there enabled its members, in less than six weeks, to bring to light all the elements needed for the fine restoration of the building published by Blouet in his magnificent work on the expedition in the Morea. In addition to this, and in spite of the want of skill

* Laloux and Monceaux, *Olympie*, p. 43.

shown by Dubois in the management of the excavations, important pieces of sculpture were found and added to the collection in the Louvre, especially the fine metope of Hercules and the Cretan Bull, which is one of the

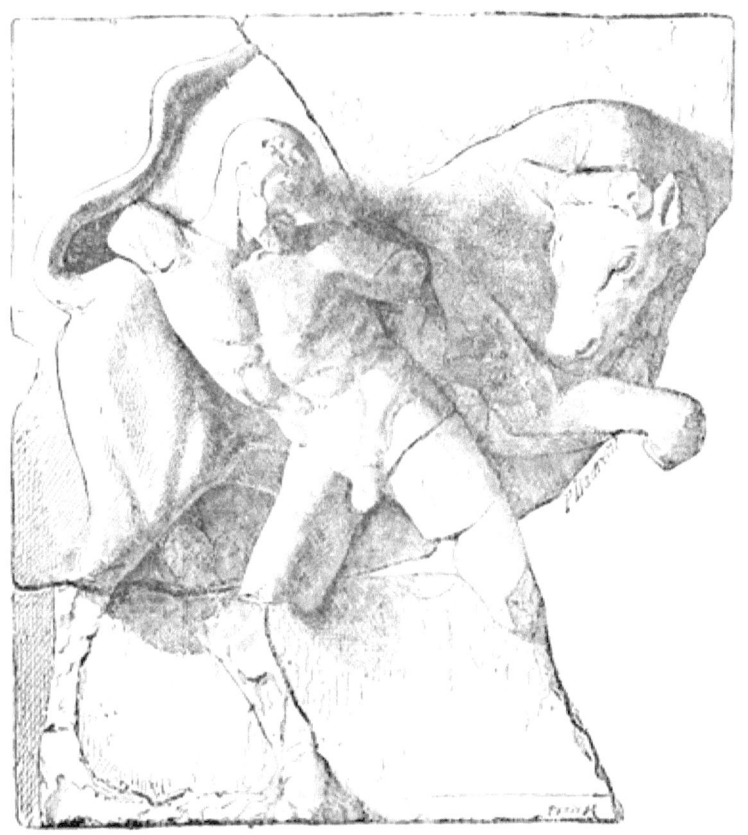

Hercules and the Cretan Bull (Louvre).

most admirable works of ancient sculpture. In spite of this successful beginning the work was soon brought to a close; it has been said that the jealousy of the Greeks had something to do with this interruption, and that Capo d'Istria, then at the head of the Provisional

Government, prohibited explorations which were robbing Greece of a part of its national glory. The explanation is perhaps simpler and more prosaic; the summer's heat no doubt compelled the expedition to suspend its operations, and its return to France prevented their being resumed in the autumn.

However, this may be, these excavations at least proved that the sculptural decoration of the temple of Zeus had not entirely perished, and that some works of the masters who had contributed to its splendour might be found among the ruins. Accordingly, all who visited Olympia strongly urged the resumption of excavations which could not fail to yield important results. Beulé on more than one occasion expressed a hope that the French Government would organise a new expedition, but the wish remained unheard. The distinguished professor of the University of Berlin, Ernst Curtius, was destined to be more fortunate. Entrusted in 1844, as it fortunately happened, with the office of tutor to the Prince of Prussia, he was able as early as 1852, in a speech which is still famous, to recommend the undertaking to the favourable notice of his royal pupil; and the young prince, who already showed signs of the liberal and cultivated mind which in later years distinguished the noble and unfortunate Frederick III., promised that when the time should come he would favour the enterprise. The time came after the war of 1870, when the newly-established German Empire sought to add the honours of peace to those of war, and the Imperial Treasury being well filled, Curtius at last obtained the necessary funds for the exploration of Olympia; and a convention concluded with the Greek government on April 25th, 1874, authorised Germany to begin the excavations. The Imperial government promised to defray the cost of the undertaking, while Greece was to retain whatever was found in the course of the work.

The German Reichstag voted an initial grant of £8000 without hesitation, but the Greeks were not so eager to ratify the treaty. The Archæological Society of Athens protested; the Chamber and public opinion were distrustful. No one could believe that the enterprise, undertaken solely in the interests of science, would not add one marble fragment to the Berlin museums, and even in 1876, a year after the beginning of the excavations, a Greek of high standing cautiously congratulated a German savant on the rare skill with which his countrymen exported the marbles of Olympia from Greece. The suspicion was excusable; the memory of Lord Elgin carrying off the sculptures of the Parthenon is one of those things which a nation does not forget. Consequently a long struggle had to be carried on with the stubborn patriots, who thought it a disgrace to leave to strangers the task of bringing to light the masterpieces of ancient Greece, but in spite of this opposition the convention was finally ratified on November 11th, 1875; the excavations had commenced a month before.

They were carried on without interruption during six seasons, until May 20th, 1881, thanks to the liberality of the German government and the generosity of the Crown Prince, who more than once drew upon his private purse, thanks, in short, to the thirty or forty thousand pounds which Germany was thus willing to spend upon a purely scientific enterprise, from which she did not acquire a single marble for her museums. During six successive years, from October to May, three hundred workmen were employed in this colossal task, which consisted in clearing, to a depth of from sixteen to twenty-three feet, not only the sacred precincts surrounding the temple—the Altis, strictly speaking, a vast rectangle more than six hundred and fifty feet long and about five hundred and seventy-five feet broad,—but also the numerous buildings which surrounded the sanctuary and were devoted to the administration of the temple, to the games which formed the glory of Olympia, and to the accommodation of the strangers attracted by the renown of its festival. "One hundred and thirty marble statues or reliefs, thirteen thousand objects in bronze, six thousand coins, four hundred inscriptions, a thousand objects in terra-cotta, forty buildings: such was the marvellous spoil won by the explorers."* It is hardly necessary to explain the importance and the absorbing interest for the history of Greek architecture and sculpture possessed by these memorable excavations. At the present time all these artistic treasures are collected in the galleries of a museum built at Olympia by the liberality of a generous Greek gentleman, M. Zingros, and thanks to this museum we can to-day, like the pious pilgrims of old time flocking to the temple of Zeus, survey the collection of remains, perhaps without a parallel in the world, which once formed part of the ancient Olympia.

Thanks to these recent excavations, the great sacred city springs up again beneath our eyes as Pausanias

---

* Laloux and Monceaux, *La Restauration d'Olympie.* p. 47.

saw and described it, and we can not only discover the exact plan of its buildings, but can restore these ruined temples, and recover Olympia as it was in the bright days of its splendour and in the freshness and youth of its monuments.

II.

Nature wonderfully adapted Olympia to its future destiny, by giving to the plain of Elis an incomparable charm. On the northern side, the mountains of Achaia and the lofty table-lands of Arcadia protected it against the cold north winds; on the south the solid mass of Messenia tempered the heat of the burning southern breezes; it was only the soft, damp western wind which could penetrate through the opening towards the sea into the river valley where Olympia lay. Thus the summits of the hills which encircle the plain, and the well-watered valley lying in its depths, were clothed with a rich vegetation, especially at the spot where, at the foot of Mount Cronion, the Cladeus united its waters with those of the Alpheus. On the heights the pine-woods added their note of dark green, while in the plain below a dense grove enveloped the shrine in its shadow, and gave to the precincts of Olympia the name of Altis, which means a sacred wood. Beneath the shelter of the avenues of plane trees there rose a crowd of temples and monuments, while clumps of silvery poplars overshadowed the tomb of Pelops, branching olives grew round the stadium and the hippodrome; and all around, on the sides of the hills, the vine, the olive, and the myrtle mingled their foliage, and the meadows in the valley, carpeted in the springtime with bright-coloured anemones, made the Olympian shrine a fresh and verdant retreat. An Athenian orator says that Olympia lies in the most beautiful district of all Greece, and certainly an inhabitant of the dry and barren Attica could not fail to be transported with admiration when he saw the charming and verdant valley of the Alpheus. We naturally find more originality and more fascination in the severe beauty of the Attic landscape, in the clear, harmonious lines of Parnes and Pentelicus, standing out against the luminous evening sky, in those bare and lofty mountains where the olive alone breaks the monotony with the pale foliage which can assume such rich and wonderful colours in the sunlight; and the pleasant valley of Olympia, with its chastened and simple beauty, its peaceful, smiling charms, has less power to move us with pleasure or surprise. To the ancient Greeks however, who like the Mussulmans of our own day were lovers of flowers and greenery and shade, its groves and running waters formed a fitting background for the temples and statues erected within its precincts, and for the festival which was celebrated there.

Let us now imagine that on one of these days of solemn rejoicing we are arriving at Olympia, like the pilgrims of olden time, by some one of the seven great roads which connected the shrine with the rest of Greece—by the most famous of them for example, the Sacred Way which ran from Elis to Olympia over the plain and met the road from the sea and the port on the banks of the Alpheus. " It was a charming avenue, lined with shrines, statues, and tombs, and with groves sweet with flowers, which led like a roadway through a park to the right bank of the Cladeus." Let us cross the torrent; in the angle which it forms with the Alpheus at the foot of Mount Cronion, which overhangs it on the north, and on whose slopes some of its buildings rise one above the other, we shall find the sacred city of Olympia. It consists of two divisions; the one is the consecrated enclosure of the Altis, encompassed on three sides by white walls; the other consists of the numerous buildings erected around it. The former is the domain of the gods—the latter, the accommodation provided for the amusement or the convenience of their worshippers.

Several gateways gave access to the Altis. We will enter by one of the three constructed in the western wall, of which the middle was a postern, while those at the extremities were more imposing in appearance. That on the south-west was the triumphal gateway through which sacred processions entered the sanctuary. It leads us to the terraces surrounding the principal building within the temenos, the most famous of the wonders of Olympia, the great temple of Zeus.* It was a fine building of the Doric order, nearly contemporary with the Parthenon, 210 feet long and 65 feet high, with six columns at the front and back; its pediments were decorated with sculptures ascribed by tradition to Paeonius and Alcamenes, and in the cella the famous gold and ivory statue from the hand of Phidias represented the Olympian Zeus in his serene and sovereign majesty.

In our rapid passage through the Altis we will not

* See Laloux and Monceaux, *Restauration d'Olympie*.

pause before the multitude of statues and monuments covering the open space in front of the temple, and multiplying at every step the image of the king of heaven, but will make our way at once to the great altar in the shape of an ellipse which occupied the centre of the Altis, and on which, according to tradition, was offered the first sacrifice in honour of Zeus. It consisted of two parts, one above the other: first a large circular platform on which the victims were slain, and then a second terrace, entirely composed of the ashes of the victims, on which rose the altar where the flesh was burned. Here the prophets of Zeus, transported by a sacred frenzy, foretold the future; here was the true centre of the cultus of Olympia. Its height of more than twenty feet, which continued to increase every year, bore witness at once to the antiquity and to the fervency of the worship paid to the supreme deity of the Altis.

If from this central spot the pilgrim cast his glance over the different buildings within the temenos, he would first find, to the north of the temple of Zeus, the Pelopium or tomb of Pelops, a lofty mound heaped up over the ashes of the hero, before the enclosure of which stood a portico open towards the west. Then came the ancient sanctuary of Hera, a very early temple of the Doric order, full of offerings, statues, and curiosities of every kind, among which were a toy once belonging to Hippodamia, and the celebrated cedar chest, richly inlaid with gold and ivory, in which Cypselus tyrant of Corinth had been hidden by his mother. Here, too, were the magnificent table on which the victors' crowns were placed, a masterpiece of the

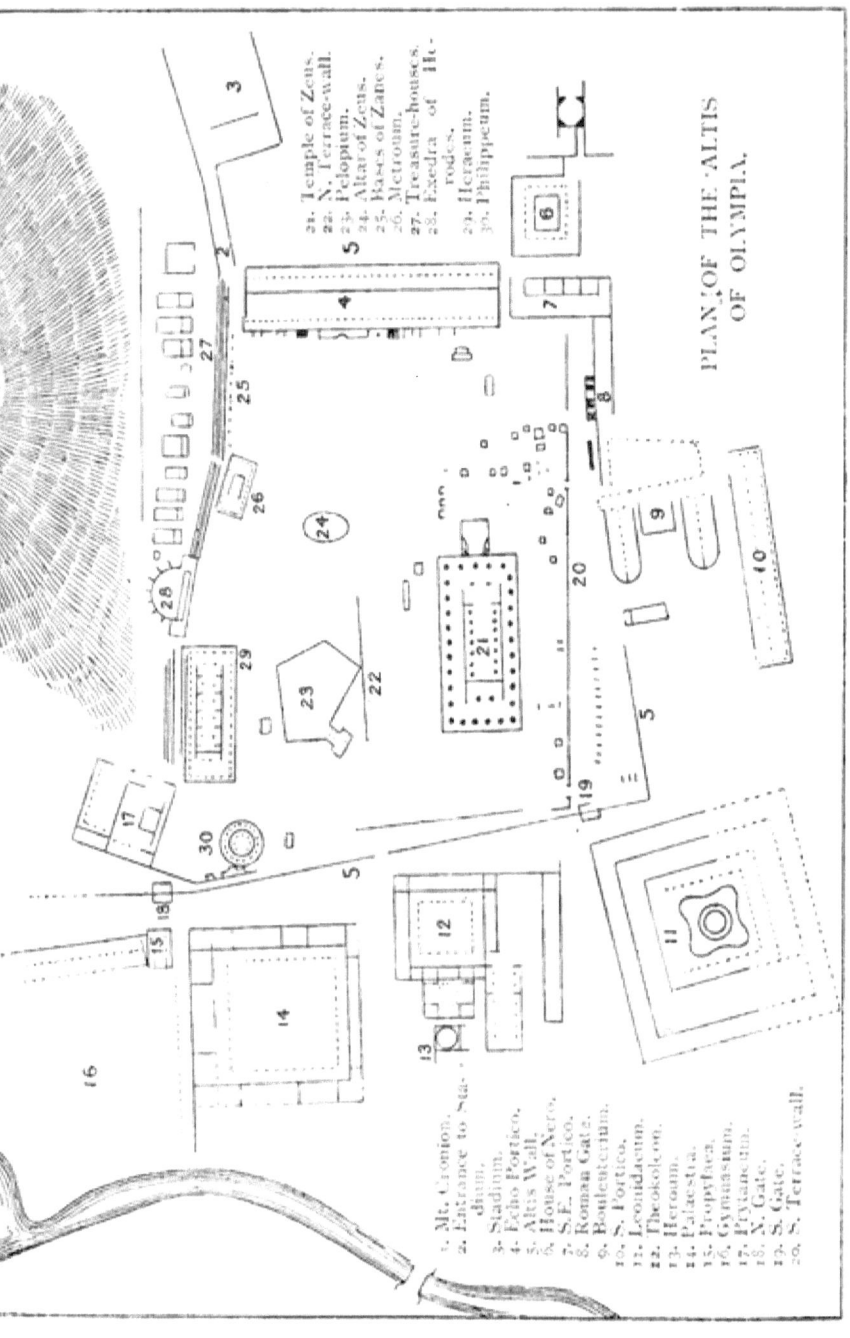

industrial art of the Greeks, and that precious statue by Praxiteles, Hermes carrying the infant Dionysus, which we have had the good fortune to recover. Farther on was a third temple, also of the Doric order, dedicated to the mother of the gods and called the Metroum. Lastly, to the west of the Heraeum stood two other buildings; the circular edifice built by Philip of Macedon after the battle of Chaeronaea, a graceful Ionic building ornamented with gold and ivory statues of the founder and his son Alexander, and the Prytaneum, where solemn banquets were held.

Above these buildings a terrace raised to a height of from ten to thirteen feet, and ascended by broad flights of steps, supported a series of buildings ranged on the slope of Mount Cronion; these were the thirteen treasuries, erected by a certain number of towns in order to receive the valuable offerings they had dedicated to Zeus. Here, too, visitors found numberless curiosities to admire; in the treasury of Sicyon they were shown the sword of Pelops and the two great bronze coffers given by the tyrant Myron; in that of Syracuse were the presents sent by Gelo in memory of his victory over the Carthaginians. Then came the treasuries of other towns of Magna Graecia and Sicily, of Metapontum and Sybaris, of Selinus and Gela, and that of Megara, from which important fragments of decorative sculptures have survived. Lastly, to the west of this series of buildings was a remarkable construction of the Roman period, a semicircular edifice or exedra, in front of which a large basin received the waters of an aqueduct, a very valuable present made to Olympia by the rhetorician Herodes Atticus. There was no spring in the neighbourhood of the Altis, and in summer the trees within the sacred precincts as well as the pilgrims who flocked to the festival, ran great risk of dying for want of water, and consequently the Elean engineers had been early compelled to build cisterns and lay down pipes, in order to supply Olympia during the dry season and keep it cool and fresh. Herodes Atticus resolved to do more; he turned from its course an affluent of the Alpheus, and led it by a great aqueduct to the side of Mount Cronion, whence its waters were distributed over all the Altis. Twenty-one marble statues were set up round the exedra, and on the building itself was inscribed the name of Regilla, the wife of Herodes, in whose name it was dedicated.

On the east a long portico adjoined the Altis wall, which was called the Poecile or Echo portico, on account of a celebrated echo which repeated seven times any word uttered there. From this vast gallery, which afforded shelter from the rays of the sun and the inclemency of the weather, there was an admirable view over the whole of the sacred place, and pilgrims crowded there during the festival. Before the Echo portico stretched the agora, covered with statues, altars and ex-votos. Here a long tribune, parallel to the portico, was probably reserved for the magistrates, and possibly was also used for those public readings at which the most illustrious writers of Greece, orators, philosophers and historians, recited their works to their assembled countrymen. To the south of the Echo portico another gallery, the South-East portico, extended in front of the house of the Hellanodicae, in which distinguished guests were sometimes lodged, and where Nero took up his abode when he visited Olympia.

To these buildings must be added the multitude of altars contained within the sacred enclosure, and visited every month by the train of priests. Pausanias reckons sixty or eighty of them, each of which required a sacrifice at least once a month. There were also the relics piously preserved in the sanctuary, such as the olive-tree planted by Hercules, and the old wooden pillar from the house of Oenomaus. There were also a multitude of statues erected to the gods, and especially to Zeus, by the different states of Greece, among which were the statue dedicated by the victors of Plataea, bearing on its base the name of the towns which had taken part in the battle, and the colossal figure, nearly thirty feet high, dedicated by the Eleans in memory of their victory over Arcadia. Other monuments recalled instances of perjury committed by the athletes during the games; thus, at the foot of the terrace on which the treasuries were built, a line of statues of Zeus, called Zanes, had been erected with the proceeds of the fines inflicted upon the guilty. Beside all these we must picture to ourselves the multitude of offerings, of groups and ex-votos, which in some places, as for example in the space before the temple of Zeus, covered the Altis with a crowd of statues. On one side, above the western terraces, were to be seen a band of thirty-five children in bronze, consecrated by the town of Messina, and farther away other children, executed in stone, by Calamis, an offering to Zeus from the people of Agrigentum. Further on were the group of Onatas, representing the Greek heroes drawing lots for the honour of meeting Hector in single combat, the colossal statue of Hercules, sixteen feet high, the splendid offerings of the Sicilian Micythus, guardian to the children of Anaxilaus of Rhegium, the bronze bull of the Eretrians, and above all the Victory of Paeonius, dedicated by the Messenians of Naupactus, which has come down to us. To all these we must add the innumerable statues of athletes, and representations of the victorious chariots and horses; we must also remember, if we would appreciate this incomparable museum as it deserves, that these bronzes and marbles bore the signatures of the most famous artists of Greece.

Outside the Altis, numerous buildings were set apart for the celebration of the games and for the accommodation of the guests and the various officers in the temple service. On the east was the stadium, which extended for a distance of nearly two hundred and twenty yards at the foot of Mount Cronion, and was connected with the sacred precincts by a covered way, through which the procession of judges and athletes entered the course. In the same direction, between the stadium and the banks of the Alpheus, lay the hippodrome, which was rather more than eight hundred and forty yards in length; it has, however, been almost entirely washed away by the inundations of the river. To the south of the sanctuary was built the Bouleuterium, where the Olympian senate sat during the games—a curious building consisting of a square court flanked by two wings, each ending in an apse; to the south of this again there stretched a vast promenade, the view from which extended to the banks of the Alpheus, over a meadow covered at the time of the games by an immense multitude, and filled with the noise of the secular pastimes inseparable from the religious festival. Close to the Bouleuterium was a great triumphal arch with three gateways, constructed in the south wall of the Altis during the Roman period. The palace of the Senate had, however, always from the beginning been in direct communication with the sacred precincts.

Lastly, to the west along the course of the Cladeus, whose stream was curbed and guided by strong embankments, stood the great gymnasium, surrounded by porticoes one of which measured two hundred and thirty yards in length; and farther on the palaestra, where competitors in the games completed their course of training. To the south of this building lay the dwellings of the priests or Theokoleon, a kind of monastery, in which a number of cells opened into a rectangular courtyard, and by the side of which was a chapel or Heroum, consecrated to the legendary ancestors of the priestly family of Olympia. Then came the workshop of Phidias, a rectangular building of the same shape and nearly the same size as the cella of the temple of Zeus; lastly, to the south-west of the Altis lay the Leonidaeum, a vast building surrounded by porticoes, where distinguished guests, such as the Roman governor of Achaia, were entertained with sumptuous hospitality at the time of the games. It was in fact necessary at these solemn assemblies to take thought for the accommodation, if not of the throng of pilgrims who encamped in the plain, at least for that of the important personages, the princes and statesmen, and the heads of the splendid theoriai which honoured the Olympic games by their presence. These guests were lodged in the Leonidaeum, close by the state entrance to the Altis, in the finest part of Olympia, where they might enjoy an admirable view of the marvellous scene which the sacred city presented.

III.

Here, every four years, at the time of the great festival, the Greeks assembled to celebrate the solemn games which for some days made Olympia the true capital of the Hellenic world. We know how keen was the interest which these competitions excited, how great an importance was ascribed by the ancients to physical exercises, and how supreme was the glory which victories in the arena conferred upon the conqueror.

"As water," says Pindar, "is the best of the elements, as gold is the most precious among the treasures of mortals, as the light of the sun surpasses everything in brightness and warmth, so there is no victory greater than that of Olympia." We must admit that among these contests in the Stadium of Olympia, over which all the most distinguished men of Greece were transported with enthusiasm, more than one seems to us remarkably brutal, and some of the most illustrious athletic feats very closely recall the exploits of prize-fighters. Yet at the present time, when physical exercises are regaining a recognised place in our schools, when schoolboy games interest the public and absorb their attention, and physical regeneration seems to be the means and complement of moral regeneration, it may not be without interest to study the means by which the ancients maintained the vigour and developed the grace and activity of the body, and to notice how, together with a few professional wrestlers, little better than prize-fighters, the ancient system of gymnastics also developed and produced, in the plains of Olympia, a far greater number of good soldiers and brave citizens.

The Olympian games had a very modest beginning. At first, according to the tradition of its founders, a single prize was offered and a single contest arranged, the foot-race, which consisted in running once the length of the stadium. Gradually other contests were added in order to increase the attraction of the festival and to draw a greater number of spectators by varying the games and rendering them more complicated. To the different kinds of races successively introduced the pentathlum was added, which was a combination of five different trials of strength and skill, running, hurling the discus, throwing the spear, leaping and wrestling; then came fighting with the cestus, or boxing, and a little later, in 680 B.C., contests in the hippodrome, chariot-races, and shortly after horse-races; finally a combination of wrestling and boxing was devised, called the pancratium. Henceforth the number of contests was complete. It is true that some novelties were introduced in details: races of mules and of colts were added to the horse-racing, two-horse as well as four-horse chariots were admitted to the hippodrome, contests of boys were added similar to those of men; still nothing distinctive was added to the Olympian programme. These additions to the contests remarkably increased the splendour of the festival. At first it lasted only a single day; but as the list of contests lengthened and competitors grew more numerous, this one day was no longer sufficient, and on one occasion, in 472 B.C., the foot-races and the pentathlum lasted so long that evening was falling before it came to the turn of the pancratium, and the wrestling went on far into into the night. The duration of the festival had to be prolonged, and from this time forward the Olympic games lasted five whole days. They must be studied in their final shape, in the full brilliancy of their splendour, in order to understand the irresistible attraction which the approach of the quadrennial Olympic festival exercised over all Greece.

The supreme control over the games belonged to the city of Elis, which delegated the superintendence to ten magistrates, called Hellanodicae, or judges of the Hellenes. These officers, appointed afresh each Olympiad, were entrusted with the organisation of the festival, the direction of the games, and the award of the prizes.

In order to fit themselves for their difficult task, they found it necessary to prepare themselves for it some time before the festival, and in fact they entered upon office ten months before the games began, while in order to render themselves fully equal to their delicate and difficult functions, they presided over the games in the gymnasium of Elis, and watched the long training which horses and men underwent, in order thus to complete by long and constant practice the theoretical knowledge they possessed in these obscure matters. At the same time they received the entries of candidates desirous of competing, and decided by the light of the traditional regulations of Olympia whether the competitors fulfilled the required conditions.

It was not permitted to every one to contend in the Olympic games. For a long time it was necessary for competitors to prove that they were of pure Hellenic blood, and the story is well known of the king of Macedonia who had to bring forward satisfactory evidence to show that he was not a barbarian. Still this condition was liberally interpreted, for even the Romans were allowed to compete, as being near kindred of the Greeks. It was also necessary to be of free birth, to possess a clear conscience, and to be at peace with gods and men. Any man who had committed a crime, or was guilty of impiety or sacrilege, was mercilessly rejected. The Spartans were for a long time excluded, on account of having violated the sacred truce, and the Athenians met with equally stern treatment for refusing to pay a fine to Zeus. As the oracles moreover were always inclined to side with one another and to render each other mutual support, Delphi refused its services and the Pythia was dumb until Olympia received satisfaction.

It was naturally an advantage to be rich if one wished to take part in the games; the training the competitors had to undergo, the cost of the journey, and the numberless expenses incurred in the course of the festival, amounted to a considerable sum. In this way certain contests, such as the horse and chariot-races, seemed naturally more aristocratic; and men of wealth and standing were inclined to disdain the simple gymnastic exercises which were less costly, and consequently more accessible to the middle classes. They feared they might find themselves in somewhat mixed company.

Should the competitor succeed in satisfying all these conditions, he was still only at the beginning of his task. When his name had been entered, he had to undergo a special course of training lasting ten months, and to submit during this time to a particular *régime*. Finally, he had to go to Elis to spend thirty days there, in accordance with the regulations, and to practise every morning in the gymnasium in the presence of the judges, who caused the competitors to pass certain preliminary examinations, and divided them into different classes according to age and strength. Then, some time before the festival, judges and athletes, chariots and horses, repaired to Olympia by the Sacred Way, and the competitors continued their exercises in the great buildings constructed expressly for their use, the gymnasium and the palaestra, the ruins of which may still be seen along the western wall of the Altis, affording one of the earliest examples of this kind of building. Meanwhile the magistrates entrusted with the organisation and oversight of the festival made their final preparations for the necessary clearing of the course, for the accommodation of visitors, and the good order of the whole ceremony.

At last the great day arrived. The festival recurred every four years, at a fixed date—the time of the full moon following the summer solstice. It was generally at the end of June or beginning of July, and we may well believe that at this time of the year the heat was often overpowering in the deep valley of the Alpheus, and that some courage was needed to sit throughout the day in the stadium or the hippodrome under a burning sun. It was no matter. From the day on which the sacred heralds carried the official message announcing the date of the festival throughout the Greek world, and proclaimed everywhere the sacred truce, from all sides men set out to take part in the games. Generals and statesmen, who sometimes, like Themistocles and Philopoemen, met with an enthusiastic reception, philosophers and literary men like Anaxagoras and Pythagoras, Socrates and Plato, Gorgias and Demosthenes, poets like Pindar and Simonides, seers like Apollonius of Tyana—in short, all who were distinguished among their countrymen by wealth or talent —made the Olympian games their place of meeting. The cities of Greece sent solemn embassies or theoriai, resplendent in purple and gold, who vied with one another in splendour, and proudly displayed in their magnificent chariots the costly offerings they had brought for the god. The kings who took part in the horse and chariot-races, and the men of rank and wealth who came to be present at the festival, made an equally pompous display of their riches. One went to Olympia at least as much to see as to be seen. Then there was the crowd of pilgrims whom an ardent curiosity attracted to this festival, and a multitude of artists in search of lucrative orders, and of merchants and manufacturers of all kinds on the watch for bargains. A great fair was held at the same time as the religious festival; the tents stood in rows upon the plain; the river was crowded with boats bringing provisions for the multitudes encamped at the foot of the Altis, while the roads were full of innumerable flocks and herds for sacrifice, and in the meadows there stretched interminable lines of many-coloured tents and joyous canteens, and from the whole crowd assembled in the plain of Olympia there rose the sounds of gaiety and pleasure. Women alone were excluded from the festival. They might run their horses in the hippodrome, and thus win a crown, as did Cynisca, the sister of Agesilaus, whose bronze horses stood under the trees of the Altis, but they might not enter the sacred enclosure. While their husbands, their sons and their brothers were contending in the stadium, they were banished to the southern bank of the Alpheus, and heard the sound of the cheering from afar. Once only was this rule broken. Pherenice, the daughter of a celebrated Rhodian wrestler, whose family boasted that they were descended from Hercules, could not bear to leave her son while the contest was going on, and disguising herself as a man and pretending to be a teacher of gymnastics, she mingled with the groups of gymnasts. When her son was proclaimed victor, however, her feelings carried her away, and forgetful of prudence she rushed to embrace her child. In her haste her robes became disordered, and her sex was revealed. The law was explicit: every woman found within the precincts was condemned to death. Nevertheless the judges acquitted her, in recognition of the fame her family had won, but to prevent any repetition of the occurrence the masters had thenceforth to present themselves naked as well as their pupils.

On the first day of the festival, in the early morning, the games were inaugurated by solemn homage paid to the gods. An imposing sacrifice was offered to Zeus in the name of the Elean state, and throughout the day sacred embassies were crossing the Altis and offering their gifts at the shrines. Meanwhile without the enclosure, and in the Bouleuterium, the final preparations were being made for the games. All those who were to take any part in the contests—athletes, charioteers, trainers, judges—swore solemnly that they had obeyed all the regulations and had been guilty neither of impiety nor of sacrilege, and with their hands on the altar they promised to act uprightly in all the coming contests. Then the Hellanodicae divided the competitors into classes, the wrestlers were paired by lot, and starting-places were assigned to the foot-racers and chariots. The evening was spent in conversation and in various pastimes; statesmen withdrew together to settle their negotiations, friends who had met again after long parting forgot themselves in endless discourse, athletes took counsel with themselves and waited quietly, gathering strength for the morrow, while under the starlit sky the crowd of pilgrims slept in the hope of the coming festivities.

With the first rays of the rising sun the festival began. Long before this time, while Olympia was still wrapped in shadow, a confused noise told that the multitude was waking; bands of pilgrims hurried to the stadium to secure good places, and long before sunrise the high banks of earth surrounding the arena, on which forty thousand people could find seats, were covered by a crowd of spectators. At the moment when the first rays of the sun fell upon the plain from the lofty summits of the Arcadian mountains, the sound of music was heard, and the official procession entered the stadium through the covered passage connecting it with the Altis. The Hellanodicae, in long purple robes, seated themselves on the platform erected near the goal, the trainers accompanied their pupils and gave them parting words of counsel, the deputies from the cities and the strangers of distinction took possession of the seats of honour reserved for them, while the competitors answered to their names and took up their appointed places.

The stadium at Olympia formed a long rectangle, 692 feet long by 105 feet wide; the track was 465 feet in length,—Hercules himself was said to have measured it with his mighty foot. Around it a sloping bank took the place of seats, and round the stadium, by the side of a narrow stone boundary which the spectator might not pass, was carried a water-channel, through which the water for the use of athletes and attendants ran into basins. The competitors disrobed under a tent at the western end, and the games began. They lasted three days, the first of which was reserved for the contests of children, and the two last for the contests of men; the games, however, were the same for both classes of competitors.

First came the foot-races, the most ancient of the contests of Olympia. The earliest of these was the single course—*stadion* or *dromos*—a test of speed which consisted in running once the length of the stadium, and was one of the favourite sights with the spectators, because by its rapid motion it displayed to more advantage than any other, beauty of contour and shapeliness of limb. Then came the double course—diaulos—in which the competitors had to run twice the length of the stadium, and the long race or dolichos. In this last, in which the competitors ran twelve times round the arena, or

14½ kilometers, it was less a question of speed than of endurance, and consequently the race resembled walking rather than running. These contests were rendered more difficult by the fact that the track, instead of being on firm and solid ground, was covered with a thick layer of fine sand into which the runner's foot sank, thus doubling the exertion.

Naturally the competition began with the long race, then came the double course, and finally the single course, in which each competitor, excited by the shouts of the crowd and the cries of his adversaries, put out his utmost strength. The pitch of excitement reached by these runners was sometimes marvellous; their speed was such that one could hardly see them pass, and they attained such a height of endurance that one victor in the long race, after arriving first at the goal, ran straight on to announce his victory at Argos, his native town, arriving there the same evening. The distance in a straight line is ninety kilometers, and there are two mountains to cross on the way. The victors were usually represented in the posture of the short race; this is the attitude of the runner Ladas, by the sculptor Myron, a statue famous in antiquity.

After this the wrestlers were called into the arena. More skill was needed for this kind of contest, as well as a special training. Brute force was in fact of less value than skill and science, a quick sure eye to follow and foresee every movement of the adversary, skill in parrying, ingenuity in thrusting, and variety of feint. Like fencing, wrestling was an art in which it was not merely a question of conquering, but of conquering with grace. As a rule the victor must have

thrown his opponent three times in such a way as to
make his shoulders touch the ground; but it was not
easy to grapple with these wrestlers, whose bodies were
rubbed with oil, and the rule of the Olympian games
allowed all sorts of wiles, such as stretching out the
leg, pulling the foot of an adversary, or leaping with
one bound on his shoulders from behind. Sometimes
wrestlers even grappled their rivals with their whole
strength, and pulled them to the ground by their own
weight,—this was the favourite stroke of the celebrated
Milo of Croton. Another kind of wrestling consisted
in continuing the struggle even when one of the combatants had fallen to the ground, which then became
an actual hand-to-hand fight, which might be carried on
in any way. Sometimes they clutched each other's
throats, or bit each other till the blood flowed. "And
their backs creaked, gripped firmly under the vigorous
hands, and sweat ran down in streams, and frequent
weals along their ribs and shoulders sprang up red
with blood, while ever they strove amain for victory"
(*Iliad* xxiii., l. 714). This is the moment represented
by the celebrated group of wrestlers in Florence. The
contest was not at an end until one of the combatants
acknowledged himself defeated.

Fighting with the cestus was an extremely cruel and
barbarous kind of wrestling of which boxing may offer
a very much softened resemblance. In this kind of
pugilism the athletes wound round their hands strips
of leather studded with nails or small plates of lead—
an equipment shown in the Wrestler of the Dresden
Museum. Thus armed, the combatants fell upon each
other, and struck the most terrible blows, coming out

of the struggle in a very much battered condition in consequence. When the Homeric heroes are making ready for this terrible contest they speak of nothing less than tearing the flesh and breaking the bones of their antagonist, and as a fact the defeated combatant goes away, trailing his limbs and spitting blood, with head hanging down, and ready to faint. Sometimes the combatants were left upon the field, or at least they went away with nose, ears, and teeth much damaged: indeed, this was so commonly the case that the monuments generally represent the victorious boxers with their ears much swollen—as, for example, the fine bronze discovered in the gymnasium at Olympia. Sometimes the unfortunate men returned

Bronze head of athlete (Olympia Museum).

from the fight quite unrecognisable. "After twenty years," says one epigram,* "Ulysses was recognised by his dog Argos; but as for you, Stratophon, after four hours' boxing you are unrecognisable not only by a dog, but even by your fellow-citizens. What do I say? Were you to look at yourself in a glass, you would exclaim with an oath, I am not Stratophon."

* Lang, Leaf, and Myers, *Iliad*, p. 471.

An ancient physician declared, indeed, that boxing was an excellent remedy for dizziness and headache, but we must confess that the treatment was somewhat drastic. The fight lasted until one of the boxers confessed his defeat, and the highest skill consisted in dexterously avoiding a blow rather than in parrying it; the greatest feat was to win without having received a single blow, and, better still, without having given one, but having tired out one's opponent so completely that he was compelled by exhaustion to give up the struggle.

The pancratium was the last contest on this day. This was a combination of wrestling and boxing, and on this account was one of the most highly considered among the contests, as it required both strength and skill; none was watched with so much interest by the spectators, and no victory was more eagerly sought after by famous athletes. The feats of the celebrated wrestlers of antiquity are well known. Their muscles, as fully developed as those of the Farnese Hercules, found no task too difficult for them. One seized a bull by the hind leg and grasped it so firmly that the animal left its hoof in his hand; another stopped, with one hand, a chariot running at full speed; Milo of Croton fastened a cord round his head and broke it by swelling the veins; Polydamas, like Hercules, met a lion and felled it to the earth. There was an inexhaustible supply of stories of this kind at Olympia, and these heroes of the stadium lost no opportunity of displaying their prowess. Most of them came to an evil end in consequence. The hands of Milo of Croton stuck fast in the cleft of a tree, and he died there, devoured by wolves; while Polydamas was crushed by the fall of a grotto which he had vainly endeavoured to hold up with his mighty hands.

The next day the games took place in the hippodrome. Unfortunately the excavations have afforded us no information about this structure, and we only know it from the description of Pausanias. It was no doubt parallel to the stadium, and was four stadia in length, while the track, properly so called, only measured two stadia—that is to say, 770 meters. It was long and narrow in shape, and was terminated on the east by a semicircular slope, on the west by the starting-place or *aphesis*, which was furnished with parallel stalls facing the course where the chariots or horses were stationed after the lots had been drawn. In the centre of the starting-place was an altar, surmounted by an eagle constructed in such a way as to rise mechanically and give the signal for the start. At the same moment the ropes fell which closed in the stalls, and when all the competitors were in line at the second starting-place a flourish of trumpets gave the signal again.

First came the race of four-horse chariots. The body of the chariot was mounted on two low wheels, two horses were harnessed to the pole, and there were two trace-horses or outriggers as well. The charioteers drove standing, and holding reins and whip. This was the most fashionable contest, the one which attracted the richest and most powerful of the Hellenes, and in which success was most eagerly desired. Among the victors may be found Cimon, Alcibiades, Gelo of Syracuse, Hiero and many other famous men. No sight was more exciting for the spectators than that of

the chariots dashing forward and striking against one
another on the course, or than the horses rearing madly
as they passed the mysterious turning-point where
lurked the demon Taraxippos—the terror of horses.
The thrilling description of these eager contests, in
which more than one of the competitors was often thrown
to the ground, should be sought in the *Iliad* or in the
famous lines of the *Electra* of Sophocles.[*] The chariot
races were succeeded by the horse-races, also of great
importance, in which the course was twelve times round
the hippodrome. The victorious horses were overwhelmed with honours, statues were erected, and
splendid tombs built for them. Sometimes even, like
Cimon's steeds, they were buried with their master in
the family grave.

As in the races of our own day, it was not the
charioteer or the rider who carried off the prize, but the
owner of the horses which ran. It did not even matter
if the jockey were unhorsed in the race, provided the
horse completed its course. For example the mare of
Pheidolas, after having thrown its rider, ran straight
on, and slackening its pace at the sound of the trumpet,
stopped of its own accord, a good first, before the
judges' stand. Pheidolas received the prize, and a
statue was set up in the Altis to the horse which had
won such a splendid victory.

When the races were over all returned to the stadium,
and the pentathlum followed. This was the most complicated of all the contests, and the most distinguished,
the one which displayed to the greatest advantage the
complete harmony of the human frame; and victors in

[*] *Electra*, 680; *Iliad* xxiii., 262.

the pentathlum were considered the most beautiful men in Greece, " for their bodies,"* says Aristotle, " are naturally capable of both strength and speed " (*Rhetoric*, Λ 5, § 11). Victory, however, was hard to gain, for five successive contests had to be undertaken—leaping, hurling the discus, throwing the spear, running and wrestling—the two last of which we have already discussed. In leaping an enormous distance had to be covered, and the competitors mounted a spring-board and sprang off, holding in their hands heavy weights called halteres, which afterwards helped them to stop short at the point they reached.* In the next contest stones were at first used, and afterwards circular discs, often ornamented with carving, which were thrown as far as possible. Many famous statues, of which the most celebrated, preserved in the Massimi palace at Rome, is a copy of the Discobolus of Myron, show the different attitudes which the athletes assumed in aiming and hurling the discus. A specimen of these quoits is preserved in the Berlin Museum. Its diameter is about eight inches, and the weight about four pounds.

Last of all came the armed race, which ended the games. In this the competitors ran twice round the stadium, bearing, it appears, in early times, helmet, shield and greaves, but in later days only a shield. This is the equipment of the statue in the Louvre known as the Borghese Gladiator, which undoubtedly represents a victorious hoplitodromos.

The last day of the games was devoted to the

---

* See Professor Gardner, *New Chapters in Greek History*, p. 295.

† For spring-board and halteres see " Pentathlon," Smith's *Dictionary of Antiquities*, and for weight of discus, Prof. Gardner, *loc. cit.* — E.R.P.

Vatican copy of Discobolus after Myron (Rome).

distribution of the prizes. These were antique in their simplicity; merely a crown of wild olive from the sacred tree planted by Hercules and a palm-branch, the symbols of strength and immortality, but they were bestowed with great solemnity. The wreaths were laid upon the gold and ivory table carved by Colotes before the temple of Zeus, and the Hellanodicae placed the crown on the head of the victor, while a herald proclaimed his name and country amidst the acclamations of the crowd.

Many material advantages, however, accompanied the victories in the stadium: the successful athletes were granted for life the right of dining in the Prytaneum of their native town, they were exempted from all taxation, and received many other tokens of the gratitude of their fellow-citizens, such as a seat of honour in the theatre and often an annuity which relieved them of all anxiety for their future. Still all this was nothing compared to the immortal glory which these simple Olympic wreaths conferred on the victors' names.

At length the festival was over, and nothing remained to be done but to give thanks to the gods. The victors sought the altars of the Altis, there to offer their sacrifices and thanksgivings, and in order to enhance the splendour of this solemn procession their parents and fellow-citizens often placed their purses at their disposal, so that it was to the music of the flute and amidst the hymns of a choir that the splendid procession wound its way through the Altis. Then there were the processions of theoriai offering their homage for the last time to the gods of Olympia, the solemn banquet in the Prytaneum, to which all the victors and the most distinguished strangers who had attended the festival were invited, and the feasts given by the generosity of the victors to their relatives, friends and countrymen—sometimes, indeed, to the whole multitude assembled at Olympia. Alcibiades gave such an entertainment, but he needed the co-operation of several of the allied towns, which were glad in this way to flatter the hero of the day, who was celebrating his good fortune with such royal generosity.

To complete the picture of this wonderful festival, we must not forget to add the occupations and amusements of every kind here offered to the pilgrims. At Olympia statesmen undertook the most important diplomatic negotiations, and they made a point of depositing copies of celebrated treaties in the Altis; here, too, famous sophists like Hippias and Gorgias lectured to the assembled multitude, and writers like Herodotus read their works aloud. Sometimes, indeed, as with Lysias and Isocrates, politics stole into these Olympic speeches. Princes who were proud of their literary skill had their verses recited here, and were more than once ruthlessly hissed. Men of science expounded their discoveries, and painters displayed their pictures. It was at Olympia that Aetion showed the painting in which he had depicted the marriage of Alexander, and that Zeuxis ostentatiously displayed the money he had earned with his brush. No honours were complete until the great festival of Zeus had set its seal upon them, and even at this spot, where physical prowess was most highly honoured, Greece kept a place for intellectual achievement.

Finally, before leaving Olympia, the victors caused their statues to be set up within the sacred enclosure. The figure represented was usually an ideal, for it was only after three victories that an athlete might have a portrait statue of himself erected. We know how greatly this custom contributed to the splendour of Olympia. A multitude of statues by the first artists of their time were gradually gathered together under the great trees in the Altis. The conquerors preferred to be represented in the moment of victory, and victors in the horse-races in especial vied with one another in luxury and splendour, and often had the statues of their charioteer and horses placed beside their own.

At the present time, out of this priceless collection only a few bases of statues and heads without bodies still survive. All the rest has perished in the overthrow of Olympia.

At length the festival was over, and the sacred city was once more left in silence. It was not, however, completely deserted. In the lonely Altis the priests continued to offer sacrifices at the altars, and in the intervals between the games a great religious ceremony more than once attracted numerous pilgrims to the sanctuary. Every five years the festival of Hera assembled the women of Elis and the neighbouring districts, and the oracle of Zeus daily attracted worshippers desirous of penetrating the future. Still the interest of the great games surpassed all others, for it was to them that Olympia owed its fame, its fortune, its splendour, and the incomparable lustre of its monuments.

IV.

None amongst these monuments was more famous than the great temple built by the people of Elis in honour of Zeus, the decoration of which had been entrusted to the most eminent artists of the fifth century. By a remarkable piece of good fortune we can reconstruct this famous building to-day as it once stood on the plain of Olympia, and that not only in the main lines of its architecture, but in all the splendour of its rich sculptural decoration. The excavations have, as it chanced, recovered for us the greater part of the marbles which once adorned the temple, and they reveal to us a school of sculpture hitherto unknown. Thanks to them, some of the great masters, on whom the critics of ancient times bestowed the warmest praise, have emerged from the darkness which surrounded them, and revealed themselves to us by works which justify their renown; thanks to them, a new light has been thrown on that period of art, still so obscure, which gave birth to the predecessors of Phidias; and we must pause for a moment before these works, and before the sanctuary itself, which surpassed in fame all the buildings of Olympia.

Pausanias tells us that towards the middle of the fifth century the Eleans defeated in battle their neighbours, the men of Pisa, who disputed their supremacy, and resolved to devote the spoil won in this campaign to the erection of a temple of Zeus which should be worthy of him. The building was no doubt begun in 470 B.C., and was well advanced in 457; in 445, as we know by certain evidence, it was entirely finished;

and the rapidity with which it was built, as well as the
elegance of the design, do great honour to the architect,
Libon, an Elean. The temple was not built of marble,
but of limestone from neighbouring quarries, covered
with stucco; it was of the Doric order, with six columns
at the front and back and thirteen at the side; it was
210 feet long, and not quite 89 feet broad, and the
upper angle of the pediment was 68 feet above the level of
the ground. In the interior two rows of Doric columns,
one above the other, supported the roof; and beneath
the colonnade above the doorways of the pronaos and
opisthodomos two series of sculptured metopes represented
the labours of Hercules. On the exterior of the
temple the metopes, on the contrary, were not carved,
and painted decorations, lines of palmettos and arabesques,
together with gilded shields, set off the whiteness
of the architrave. The main part of the sculptural
decoration was reserved for the pediments; on the east
front was represented the history of the famous chariot
race in which Oenomaus lost his life, on the west
front the battle between Centaurs and Lapiths, while
above the eastern pediment there were placed, on the
apex and at the angles, a Victory of gilded bronze and
two vases of the same metal, the work of the sculptor
Paeonius of Mende. Tradition asserted that while
local artists had built the temple and carved the metopes,
the pediments were the work of foreign sculptors; and
that, just as in later days Phidias and his pupils were
summoned to Olympia to erect a statue of Zeus worthy
of the god, so the decoration of the pediments was
entrusted to two of the most renowned masters of their
time—Alcamenes the Athenian, who carved the figures

of the western façade, and Paeonius of Mende, who executed the groups of the eastern pediment. It is these very marbles which the excavations have in part restored to us.

Pausanias has informed us which episode in the story of Oenomaus was chosen by the sculptor of the eastern pediment—the preparations for the race which was to take place between Pelops and the King. In the middle of the pediment,* says this writer, is the statue of Zeus; near him, on the right, is Oenomaus, a helmet on his head; and at his side his wife Sterope, one of the daughters of Atlas. The charioteer of Oenomaus, Myrtilus, is seated before the horses, which are four in number. Then come two men who are not named, but who, it is clear, are also entrusted by Oenomaus with the care of the horses.

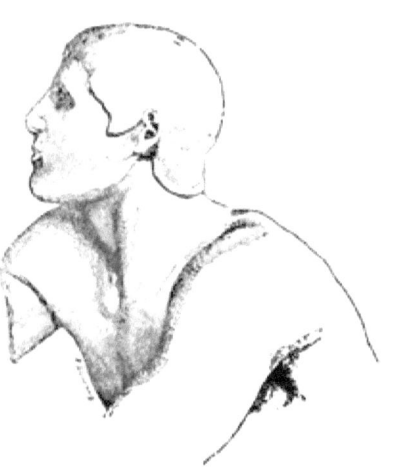

Cladeus (Olympia).

At this end of the pediment is the recumbent figure of the Cladeus. On the left, next Zeus, come Pelops and Hippodamia, then Pelops' charioteer and horses, and two men, no doubt his grooms. The tympanum contracts, and in the corner is the figure of the Alpheus.

In spite of the simplicity of this composition, which places two symmetrical groups consisting of five figures

* Pausanias, v. 10. 6.

on each side of Zeus, and fills the angle of the pediment with the two recumbent figures, it is not easy to replace the fragments of sculpture found in the excavations in the position they formerly held in the façade of the temple. The figures were all broken by falling from the height at which they stood, and those of the eastern front especially have come down to us in a very mutilated condition. The scattered fragments have had to be pieced together, and the whole composition reconstructed; and although the condition of the marble, generally left unworked at the back, furnishes some useful indications, and the height of the different figures, proportioned to the necessities of the triangular area of the pediment, helps us to fix their position very nearly, the investigations of archæologists have led them to very different conclusions, and the arrangement of the various reconstructions which have been suggested is very dissimilar.

The fragments of the eastern pediment, some of which are very noteworthy, should be studied in the museum at Olympia, or at least in the fine plates which accompany the recent work of Messrs. Laloux and Monceaux. In the centre is the nude torso, broad and grand in style, representing Zeus standing motionless and proud between two parallel groups, each of which takes up one side of the pediment. To the right of Zeus stands Oenomaus, in his right hand a spear, while the clenched left rests on his hip; of this figure we possess the finely developed bust and the mutilated head; to the left is Pelops, still young and beardless, his fair and youthful form full of grace and pliancy; and by the side of the two opponents

two figures, also contrasted with rigorous symmetry,
—Sterope in a long robe, her head slightly turned
aside, and Hippodamia in a thoughtful attitude, clad
in the heavy Doric chiton, which falls in straight,
heavy folds,—complete the central motive of the composition. Then on each side come the horses, which
effect a skilful transition between the standing figures
in the centre and
those kneeling or
lying under the
slope of the cor-
nice. Near these
must be placed the
seated youthful
figures, whether
squires or chariot-
eers; the position
of these statues
cannot be assigned
with perfect cer-
tainty, but the
natural and real-
istic grace of their
style is often ad-
mirable. Lastly,

Head of old man (Olympia).

in the left corner of the pediment is the river Alpheus,
with powerful torso and full vigorous contours raising
himself on his elbow to survey the struggle; and at the
other end, on the extreme right, two of the best preserved
and finest figures in this pediment, the Cladeus, with
slender, sinewy frame and graceful, delicate head,
raising himself on both arms, and looking before him

with intense interest, and an old man, seated between the horses and the river-god, with one elbow resting on the right leg, the left leg extended, and the lower part of the body concealed by drapery which leaves the torso bare. The head is uninjured, and remarkable for power and expression: it is bald on the crown, but on the temples and the nape of the neck it is surrounded by long curling hair; the broad forehead is deeply furrowed, the eyes look straight forward, and the lips seem to speak. In its rather coarse truthfulness the whole figure deserves particular attention; rarely in antiquity has such intense life been rendered with so much realism.

The composition of the western pediment is more complicated, and Pausanias, even in his time, found some difficulty in explaining it. In the middle of the pediment, he says,* stands Pirithous, and next him on one side Eurytion carrying off the wife of Pirithous, and Caineus bringing him aid, while on the other Theseus is defending himself against a Centaur with an axe. One of the Centaurs is carrying off a maiden, the other a beautiful boy. We see what was the subject represented here—the battle between Lapiths and Centaurs, one of the favourite motives of Attic sculptors, who reproduced it on the frieze of the Theseum, in the metopes of the Parthenon, and at the temple of Apollo at Phigalia. At Olympia, however, the scene was not so simple as Pausanias describes it; instead of the nine figures he mentions, the excavations have given us twenty or twenty-one, and it has not been easy to reconstruct the whole composition with certainty from

* Pausanias, v. 10. 8.

these fragments, which have been put together again with much trouble. Still it is evident that here, as in the eastern pediment, the groups balance one another with the most rigorous symmetry. In the centre Apollo, standing between the combatants with his right arm extended, commands the whole scene, and the broad muscular figure of the god, his youthful head with hair arranged in the archaic style, and his calm serene attitude, which is in strong contrast with the fury of the combat, give a remarkable character to his figure. On each side of the god groups of three are interlaced in a fierce struggle: on the left the king of the Centaurs has seized Deidamia, who repulses the monster with all the might of her outstretched arms, while Pirithous is hastening to her aid, and on the right a young girl is struggling to free herself from the embrace of another Centaur, whom Theseus has already wounded in the head with his hatchet.

Apollo (Olympia).

On each side of this central composition come other groups symmetrically opposed, their height diminishing with the slope of the pediment; the bodies of the young warriors are intertwined with the heavy cruppers of the Centaurs, the pliant forms of the maidens are struggling in the embraces of their ravishers, and all these figures are mingled in the most boldly varied attitudes. Finally, in the angles of the pediment women are crouching and looking with horror on the varying fortunes of the fight; and, quite at the end, a nymph is gazing, careless and unmoved, at the drama which is being acted before her eyes.*

The guides who conducted pilgrims through the Altis said that the eastern pediment was by the hand of Paeonius, and that Alcamenes was the sculptor of the western front. This tradition has been attacked and defended with much warmth since the excavation of Olympia, and in spite of this prolonged controversy it is difficult, if not impossible, to ascertain the truth in this matter; but in order to find a solution of the difficulty we must try in the first place to estimate the pediments of Olympia at their true value, and to characterise their style.

If we examine the drapery we shall find that it has been treated in an extremely superficial way. Whilst the archaic sculptors arrange their materials in heavy folds, and in the Parthenon marbles they are moulded to the form in the most marvellously pliant and harmonious way, here on the contrary the drapery is carelessly arranged, almost as if thrown on by chance, and sometimes in so strange a manner that it is almost impossible to discover a living form beneath the folds. If we study the figures themselves we shall meet with fine pieces, it is true, but by the side of these the execution in general is very shallow, the work hasty, the sculptor's hand extremely lax and feeble; we no longer find the severity and stiffness of archaic art, and we do not yet discover the graceful precision of the Attic

* See the plate in the *Restauration d'Olympie*, pp. 81-5.

school. The anatomy of these figures has not been sufficiently studied, and is sometimes incomprehensible: the muscles have no elasticity, the contours are undefined, the flesh soft and flabby; they are imperfect both in design and execution, and the artist, chiefly desirous of effect, seems to have intentionally neglected care and accuracy of modelling. As to the attitudes, they sometimes display a boldness and always a realism which is remarkable, but it would be vain to seek here for the scientific moderation found in the pose of the Aeginetan warriors, or for the incomparable nobility of the Parthenon figures. There is nothing of the ideal in the marbles of Olympia: the Cladeus is dragging himself along on his stomach, and the grooms of the eastern pediment are sitting in an easy attitude about which there is nothing academic; if the pose of the standing figures is a little more dignified, their chief value consists nevertheless in the intensity of life which sometimes animates them. Lastly, let us consider the heads, for by a good fortune which is very rare, the greater number of these marbles offer us more than a headless trunk, and whereas one only of the Parthenon figures has kept its head, at Olympia three heads from the east pediment and eleven from the western have come down to us, several among which are quite uninjured. Here again, as in the drapery and the whole figure, there are strange inequalities. Some of the faces are very archaic in appearance, the forehead is low, the skull amply developed, the hair arranged symmetrically and curled with fastidious care, sometimes even it is covered by a cap, and according to the custom of these early sculptors the brush is called in to aid the chisel. Still, by the side of this negligent work we find exquisite heads, faces shown in profile with a keen eye and an expressive mouth. Many of the female heads from the western pediment are charming in their youth and grace, and in spite of a certain coldness of expression have an indescribable air of distinction. There is, in short, a strange and unusual mingling of beauties and of defects in these pediment sculptures: archaic tradition goes hand in hand with audacious novelties; together with some figures of astonishing beauty and skill there are abundant marks of carelessness, inaccuracy, clumsiness and ignorance; and, what is still more remarkable, the same characteristics are to be found in both pediments—hasty and unfinished work (for pains have scarcely been taken to block out the backs of the statues) together with shallowness and feebleness of execution—and at the first glance it

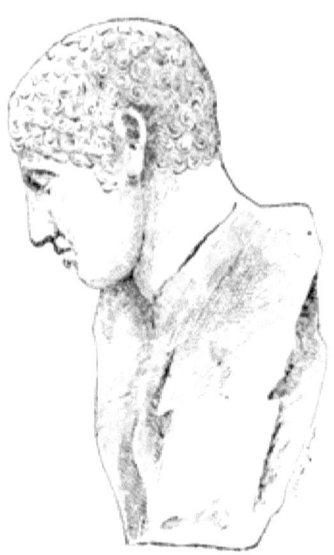

Head of a Lapith (Olympia).

Head of a maiden (Olympia).

would seem that all these marbles have sprung from the same school and are the work of the same hands.

Still, if we consider the composition, which is so superior to the workmanship in both pediments, we shall find a marked difference between them, in spite of the fact that the groups are arranged in both with rigorous and exaggerated symmetry. In the eastern pediment the figures are almost, as it were, thrown down by chance, without any link connecting them in a common action, and stand in a row, as has been well said, like a set of skittles. The thoroughly dramatic composition of the western pediment, the bold variety, the life and passion of which it is full, are in strong contrast with the stiffness and dryness of the other; we feel that the artist was ambitious of reproducing action in marble, and desired to arrange his figures in skilful combinations; and in spite of some defects the work has not fallen very far short of his conception. Consequently our analysis leads us at last to this strange conclusion: the two pediments cannot have been designed by the same master, and yet from their style they seem to be by the same hand.

Are we to believe, as an explanation of this identity in the workmanship, this inequality between design and execution, these defects which could scarcely be found in the work of a great artist, that the Athenian sculptors only furnished the designs, and that local workmen executed the sculptures from their drawings? Such a division of the work was not unusual, and Phidias himself was satisfied with designing the Parthenon frieze.

Or are we, on the contrary, to endeavour to discover

some striking resemblances between the figures of the east pediment and the Victory signed with Paeonius' name, and to find some traces in the western one of the refined and aristocratic tastes which characterised the style of Alcamenes ? Must we contrast the more graceful and delicate proportions of the figures on the western front, their more precise and delicate modelling, with the vigour and power, the truth and realism of the eastern façade, and in order to save the traditional account at any price, and to preserve the paternity and the credit of these statues for Paeonius and Alcamenes, must we account for the defects in their execution by supposing that these decorative works, placed at a height of sixty feet or more above the ground, were made to be seen from afar, to strike the eye boldly, and that at this distance precision of detail and finished execution were of little importance ?

It is certainly tempting to believe we have recovered the authentic work of two great masters hitherto unknown to us, to show these artists, contemporaries and rivals of Phidias but independent of his influence, working at Olympia before the arrival of the great Athenian sculptor, and paving the way for him by their works, which are earlier in date than the Parthenon marbles. It is tempting to be able to interpose between the Aeginetan sculptor and the famous Attic master these two predecessors so different in their style : the one, Paeonius, full of spirit and fire, making decorative effect his first thought, striking hard, even roughly, in order to be understood by the multitude, a kind of Michael Angelo of antiquity, as has been said, with the energy and animation, the truth and realism, which do

not shrink from a certain amount of coarseness; while
the other, Alcamenes, whose masterpiece, the Aphrodite
of the Gardens, was famous throughout antiquity * for
the graceful proportions of the hands and wrists and
the delicate contour of the face, was more refined and
aristocratic, more scientific than Paeonius, but less
powerful, more graceful and delicate but less vigorous
and striking; the charm of his works, however, was
perhaps more penetrating and more original. Both
were the heralds and forerunners of Phidias, who was
to unite the Attic grace of the one with the commanding
simplicity and majestic power which the other derived
from the robust school of the Peloponnesian masters.

Unfortunately, direct proof is wanting for the support
of these attractive hypotheses; our acquaintance with
ancient times is so fragmentary and incomplete that
whenever we try to fix their date, the most illustrious
names fluctuate between limits of the utmost uncertainty;
while on the other hand the works themselves, even to
competent judges, are to so great an extent tinged by
the observer's personality that each sees them with a
different eye. If to these difficulties we add the fact
that the marbles of Olympia have at the outset entirely
disconcerted all who have studied them by the unexpected qualities they reveal, we shall understand that
this surprise led to the greatest diversity of opinion.

"If," as Rayet well says,† "we picture to ourselves
a huntsman on the track of a deer, who unexpectedly
starts a wild boar, we shall understand the confusion
to which these discoveries have given rise."

* Lucian, *Imagines*, p. 6.
† *Études d'Archéologie et d'Art*, p. 59.

While some see in these marbles the works of a particular school originating in northern Greece (this is the opinion of one of the most competent judges *), others trace in them with equal certainty the purest traditions of Attic art; and while some detect, at least in the western pediment, striking analogies with the metopes of the Parthenon, others declare, on the contrary, that these sculptures are much earlier than the Athenian marbles, and entirely independent of the influence of Phidias. Some see in these pediment sculptures original works of Paeonius and Alcamenes, others would only leave these artists the credit of having designed them, and transfer to the members of a local school the responsibility of having carried out the work from their rough models; some are struck by the traces of archaism which these marbles still preserve, while others notice especially the characteristic signs of a new movement in art. Amidst this conflict of opinion it is singularly presumptuous to assert any conclusion; but at least we must not allow the works themselves to be lost sight of beneath the flood of criticism. They deserve an important place in the history of Greek sculpture, and to whatever school the nameless artists who created them may belong, they ought to take high rank among the immediate predecessors of Phidias.

The excavations of Olympia have given us an authentic work by the sculptor to whom tradition attributed the eastern pediment. In 424 B.C. the Messenians of Naupactus set up a statue of Victory on a lofty pedestal before the temple, in memory of the glorious

---

* Brunn, *Sitz. Ber. d. Kön. Bayr. Akad. d. Wiss.*: "Die Sculpturen von Olympia," 1877.

success they had gained at Sphacteria over their

Victory of Paeonius (Olympia).

hereditary enemies the Spartans; and on the base of
this monument, which has fortunately been preserved,

we read at the end of the dedicatory inscription, "Paeonius made this statue, and carried off the prize for the acroteria placed upon the temple."[*] Here there is no room for doubt, and this signed and original work reveals to us better than the pediments the great and almost unknown master who worked at Olympia.

The goddess is represented flying in mid-air, an eagle hovers beneath her feet, and her whole body is thrown forward in glorious motion; the left foot scarcely touches the pedestal, while the right still presses the marble, which was formerly painted blue, and represents the space through which the Victory is taking her flight. All the outlines of the body are visible under the fine clinging material of her robes blown by the wind, and the long Doric chiton, leaving the left leg and shoulder uncovered, swells out behind in harmonious folds. Originally her ample outspread wings, and a wide mantle floating on the breeze, supported the statue and restored its balance; originally, too, the left arm was raised, and gave the goddess a still prouder attitude, and the head, of which unfortunately only the back remains, completed the effect of this wonderful figure. Still, mutilated as it is, it is a splendid work, and we cannot sufficiently praise the life and picturesqueness of the design, its boldness of imagination and unequalled spirit, its grace and elegance, and its deep feeling for nature. When we compare this Victory with the figures of the eastern pediment, it seems almost impossible to attribute such dissimilar works to the same master. It is true that, if we wish at all costs to cling to tradition, we may notice in both the same

[*] Loewy, *Inschriften Griechische Bildhauer*, No. 59.

realistic tendency sometimes running to excess, and we may remark that an isolated statue, intended to be looked at close at hand, required more finished execution than the decorative pediment sculptures; lastly, we might recall the fact that twenty years or more had passed since the time when Paeonius was at work upon the latter, and that during so lengthened an interval a great artist could not stand still. Certainly, if we could accept this hypothesis, it would be interesting to study in these marbles the gradual progress and artistic development of a famous master, and to show this proud and independent sculptor yielding, in spite of himself, to the influence of the great movement which was spreading all over Greece, and adding greater nicety of detail, more finished execution, and more graceful pliancy, to his natural boldness and strength and to the traditional qualities of his school. The work itself, however, apart from these conjectures, is sufficient to render its maker famous; it cannot, of course, be compared with the Parthenon marbles, but the artist who has made it deserves to be remembered. It seems, too, to have exercised a lasting influence on after ages. There is more than one feature in the marble of Olympia which recalls the fine Victory of Samothrace in the Louvre, and makes us think of the sculptures of Halicarnassus and Xanthus; there is the same love of violent motion, of draperies flying in the wind, and of expressive and dramatic attitudes; there is the same *élan*, the same intensity of life; and if it is true that all these works are connected with the school of Scopas, it is from Paeonius, more than from any other man, that the Parian sculptor traces his descent.

Besides the pediment sculptures, twelve metopes, placed under the colonnade of the temple, decorated the frieze of the cella. These bas-reliefs represented the labours of Hercules; important fragments of two among them were discovered in 1829 by the French expedition to the Morea; and the Germans, in the course of their excavations, have brought to light numerous remains of the other metopes, so that it is now possible to gain an exact idea of almost all. Without entering here into a useless enumeration, it will be sufficient to mention those which are sufficiently well preserved to be of interest for the history of Greek art. In the first rank come the two metopes of the Louvre: Hercules taming the Cretan Bull, undoubtedly the finest of the whole series; and Hercules victorious over the Birds of Lake Stymphalus, which has been very happily completed by the German discoveries. Then come the fight between Hercules and the triple-bodied monster Geryon, and the cleansing of the Augean stables, in which Athena, her helmet on her head, is standing behind the hero, who, in a very realistic attitude, seems to be sweeping away the accumulated filth before him with great strokes of the broom. Lastly, there is the metope showing Hercules demanding from Atlas the golden apples of the Hesperides. The legend is well known. In order to gain his end and obtain the wondrous fruit, Hercules asked Atlas to gather them for him in the famous garden, while in exchange he offered the giant to relieve him for a moment in the arduous task of supporting the world upon his shoulders. The Olympian metope represents the moment when Atlas, who has accepted the offer, is bringing back the golden apples to the hero. In the centre is the nude figure of Hercules, supporting with both uplifted arms the heavens, which rest upon his head; in front of him is Atlas, holding in each hand three apples, and to the left one of the Atlantides, wearing the heavy Doric chiton. The sculptor of Olympia did not tell the end of the story. When Atlas had once laid down his burden, he found the liberty he had regained so sweet that he proposed to leave the world on the shoulders of Hercules and to take the golden apples himself to Eurystheus. This, however, was not to the liking of our hero, who was beginning to find the weight oppressive, and, fortunately for himself, on this occasion he was as crafty as he was strong. Consequently he made a pretence of accepting the proposal, and only asked as a favour that he might make a comfortable cushion to support the burden, and that meanwhile the giant would take up his accustomed post for a moment. Atlas imprudently consented, and we may imagine that the hero, once the heavens were replaced upon his shoulders, took up his apples and made all haste to depart.

Tradition has not handed down to us the names of the sculptors of these metopes, but their date at least can be determined almost exactly. These great slabs of stone, five feet three inches by almost five feet, could not have been added after the building was completed, like the pediment sculptures; they are so well fitted in between the triglyphs of the frieze that they must evidently have been placed there when the temple was built, that is to say between the years 460 and 450 B.C. Moreover we need only look at them to see that they are not yet entirely free from the influence of archaic traditions. In the parallelism of its straight, heavy folds the drapery preserves some of the stiffness of early works; on the square, bony heads, the beards and hair are simply blocked out, and painting, according to ancient custom, has been called to the aid of sculpture; the attitudes, too, preserve an immobility which is almost hieratic. In short, if we compare these metopes with those of the Theseum at Athens which represent the same subject, we find the bas-reliefs of Olympia far more archaic in style; although they are scarcely ten years earlier in date than the Athenian marbles, they are inspired by very different traditions and are quite unaffected by the influence of the Attic school. We lose ourselves in useless subtleties if we endeavour to distinguish two different series among these metopes, one belonging to a purely Peloponnesian school and the other inspired by the teaching of Paeonius, and we shall waste our time in unprofitable investigations if we try to decide how far the style of these metopes resembles that of the pediments. These reliefs are in fact undoubtedly the work of local artists whom the architect of the temple employed for the first part of the decorative work, and certainly these Peloponnesian sculptors were no despicable artists. They were already able to shake off archaic conventions sufficiently to render, by sober and correct modelling, the precise anatomy and powerful muscular system of the body. If the designs still betray a certain clumsiness and *naïveté*, still they sometimes rise to higher conceptions, and the metope in the Louvre, by the striking simplicity of the composition, the strength and sobriety of the execution, the remarkable technical skill displayed, together with the boldness and audacity of the design, is unquestionably a masterpiece. Contrasted even with the metopes of the Parthenon, which are later in date, the relief from Olympia can hold its own. These Peloponnesian masters have, it is true, something dry and archaic in their style, but, together with remarkably careful work, an unusual comprehension of the requirements of decorative art, and an astonishing devotion to the study of nature and life, they have a power and breadth which the Attic sculptors do not possess. It is in the pediment of the Parthenon that we again meet with some of the robustness which distinguishes the metope in the Louvre; and it contributes in no slight degree to the interest these marbles excite that they enable us in some measure to understand how the work of Phidias arose.

And now, since on the sacred soil of Olympia we find, as it were, an epitome of the history of Greek sculpture, we must come to the wonder of the Altis, the famous statue of the Olympian Zeus, which in the judgment of antiquity passed for the masterpiece of Phidias and the most sublime effort of Hellenic art. When the Eleans had finished building the temple of Zeus, they felt a natural desire to erect a statue to the god which should be worthy of him; and to whom should they turn for this purpose but to the famous master who at that very time was setting up an incomparable monument to Athena on the Acropolis of Athens? The sanctuary of Olympia could not be suffered to fall short of the Parthenon in any respect, so Phidias was summoned to the Peloponnesus. When the buildings and statues on the Acropolis were finished, or possibly even some years later, after the famous trial which drove the friend of Pericles from Athens (in any case later than 438 B.C.), the great sculptor came to Olympia bringing with him a group of artists who were to take part in the work, among them the painter Panaenus, a relative of Phidias, who decorated the interior of the temple of Zeus with a series of frescoes. The Athenian artists were received with much distinction, and a splendid workshop was placed at the disposal of Phidias, exactly similar in size and arrangement to the cella of the temple where the statue was to be erected, so that the master could judge beforehand what effect his statue would produce when put in its destined place.

The masterpiece of Phidias stood in an open space cut off by a partition at one end of the cella, in the full light of day, which poured down upon it through the hypaethral opening. Pausanias,[*] who saw the statue, has left us a detailed description of it. The god was represented seated on his throne, bearing in one hand his sceptre and in the other, like the Athena of the Parthenon, a winged Victory. He was robed in a mantle, leaving the shoulder and the upper part of the chest free and enfolding the lower part of the body in its drapery. The nude portions of the figure were executed in ivory, and the drapery in gold of different tints, while the mantle was inlaid with lilies and figures in enamel. The head was wreathed with sprays of olive, recalling the prize given to the Olympian victors, and the god was seated in a calm and majestic

[*] Pausanias, v. ii.

attitude—" mild and peaceful," says Dion Chrysostom, " watching over a peaceful and united Greece." For the countenance, according to a well-known story, Phidias is said to have drawn his inspiration from some lines of Homer : *

> " He said, and nodded with his shadowy brows,
> Waved on th' immortal head th' ambrosial locks,
> And all Olympus trembled at his nod."

The majestic character of this colossal work (the statue was more than forty feet high) was completed by the throne of ivory and gold, ebony and marble, on which the god was seated. It was a marvel both of artistic beauty and of mechanical skill, in which the perfection of the chasing and the sculpture vied with the richness of the materials employed. Figures in the round supported the seat, and reliefs representing, among other things, the eight kinds of games practised at Olympia, ran round the lower part; surmounting the back, and reaching higher than the head of the god, were two groups representing the Seasons and the Graces, while on the base were the divinities which formed the train of the king of the gods, and on his golden footstool, resting on lions, was sculptured the battle between Theseus and the Amazons. The whole of Greek mythology seems to have been concentrated on this gigantic statue, in which gold and ivory, painting, sculpture and chasing, combined their colours and effects so as to produce an impression of the utmost brilliancy and most harmonious richness.

Unfortunately we know but very little of the ideal type created by Phidias. The statue disappeared,

* *Iliad* i. 622-5 (Lord Derby).

either in the burning of the temple or at Constantinople, if it be true that Theodosius I. had it transported to his capital, and amongst the numerous busts of Zeus with which our museums are filled, none can claim to be an exact reproduction of the work of

Asclepius Blacas (British Museum).

Phidias. The ideal he created underwent many transformations in later times, and the recognised type of the god was changed in the hands of his successors,—so much so that we cannot tell how far these copies are removed from the original. While some faithful replicas give us a fairly accurate idea of the Athena of the Parthenon, the Otricoli head and the Verospi Zeus in the Vatican, which were long thought to be good copies of the Olympian statue, are only poor and untrustworthy imitations.*

Authentic reproductions of the masterpiece of Phidias must be sought on the coins of Elis of the age of Hadrian. In these the hair, confined by the olive-wreath, falls quite simply in separate locks on the nape of the neck, leaving the ear free. The forehead is broad and high, the strongly marked eyebrows recall the sovereign lord of gods and men, whilst the mouth and the lower part of the face show the mild and peaceful deity of whom Dion Chrysostom spoke.

It was no angry god launching his thunderbolts whom Phidias made manifest, but the father of gods and men, in whom sovereign majesty reposes in unchangeable calm and is tempered with infinite gentleness. The ancients exhausted all their words of praise in speaking of this masterpiece of the Athenian sculptor. "Go to Olympia," said Epictetus, "to see the work of Phidias; and deem it a calamity to die without having seen it." † Whoever had the happiness of beholding it, stood transfixed, as in the presence of a manifestation of the deity. "O Phidias," says a poet in the *Anthology*, "either the god came down from heaven to earth to show thee his image, or thou didst ascend to Olympus to contemplate him there." ‡ "Should any man have his mind so overburdened with care and sorrows that even

---

* For the marble head of Zeus or Asclepius from Milo, in the British Museum, see Murray's *Greek Sculpture*. —E. R. P.

† Arrian, *Epict.* i. 6, 23. ‡ *Anthol. Gr.*, ii. 208. 48.

sweet sleep forsakes him, standing before thy statue he would, methinks, forget all that there is bitter and grievous in life, such a wondrous work hast thou conceived and brought to pass, and such heavenly light and grace hath thine art bestowed upon it." * It is said that the grave and pious Paulus Æmilius, † the conqueror of Perseus, was so struck with admiration when brought face to face with this sovereign majesty, that he sacrificed to the Greek god with the same ceremony as to the Capitoline Jupiter.

The most touching, however, of all these testimonies is to my mind the beautiful and artless legend related at Olympia itself. When Phidias had completed his work, and the statue was set up on its pedestal at the farther end of the cella, he addressed a fervent prayer to the god, beseeching him that the work might find favour in his eyes, and that he would show by some clear token if in this creation of a mortal hand he recognised a representation worthy of his sovereign majesty. Then suddenly the heavens opened, the thunder pealed, and through the opening in the roof of the sanctuary there flashed a thunderbolt which pierced the temple floor. According to Pausanias, an urn in his time marked the spot where this miraculous sign had consecrated the masterpiece of Phidias.

V.

During the long centuries through which the sanctuary of Olympia existed, many other buildings arose by the side of the temple of Zeus, many famous examples of Hellenic art were set up under the great trees of the

* *Dion Chrys. Orat.*, 12, 51.   † *Livy*, xlv. 28.

Altis, and if the excavations, as must be admitted, have restored to us but a small part of these wonders, they have nevertheless taught us many lessons hitherto unknown. In particular they have thrown quite a new light on the obscure and remote origin of Greek civilisation and art, and even if this were all, they would deserve the attention of archæologists. Mycenae and Hissarlik, Delos and the Acropolis, have rendered us familiar with the long period through which ancient art was gradually developing, and have taught us to know the somewhat uncouth but powerful charm of early sculpture. The services which Olympia has rendered us are no less valuable, for these excavations have shed an unexpected light upon the architecture and sculpture of the archaic period. To the north of the temple of Zeus stood the Heraeum, one of the oldest buildings in the Altis, dating at the latest from the eighth century before our era, and we can well imagine what importance this temple, the oldest on the mainland of Greece, must have for the history of early architecture. We know to how much controversy the history of the origin and formation of the Greek orders of architecture has given rise; for while some have asserted that Hellenic temples were from the beginning built in hewn stone, others affirm that the earliest monuments of Greek architecture were constructed in wood, and that the essential features of the classical temple were derived from this primitive building. The Heraeum of Olympia has furnished striking confirmation of this latter theory. Pausanias,[*] when he visited this temple, noticed with interest in the opisthodomos the column of oak, the remains, it was

---

[*] Pausanias, v. 16.

said, of the original edifice; and the results of the excavations have fully confirmed his statement, for while, as a fact, in the case of all the other buildings in the Altis, portions of all the principal parts of the building have been found buried in the earth, here nothing remains either of the architrave or the frieze, and the reason plainly is that they were of wood. A still more significant indication, however, is given by the infinite variety of the stone columns forming the peristyle, no two of which are alike, and whose capitals exhibit all the successive types, from the heavy primitive capital of the seventh or sixth century to the pure and elegant style of the fifth century, or the stiff, cold examples of the Roman period. It is easy to explain this dissimilarity if we admit that in the beginning all the pillars were of wood, and that in the course of centuries they were successively replaced by columns of stone.

Such was the original Heraeum; with the exception of the foundation, constructed of solid hewn stone,* its walls were built of brick, its columns and entablature were of wood, a tiled roof, the oldest with which we are acquainted, covered in the building, and in the interior engaged columns along the side walls of the cella formed a series of recesses and took the place of the rows of pillars. The terra-cotta decorations, however, together with the rich polychrome ornamentation of the upper part, formed the most original part of the building.

It was, in fact, exceedingly necessary to protect the loftiest and most prominent parts of a wooden temple, such as the cornice and pediments, against the inclemency

---

* This stone plinth reached a certain height above the ground. See Denkmaeler, *Olympia*, vol. ii., p. 1103.—E. R. P.

of the weather; and for this purpose it was customary to employ terra-cotta casings, which also furnished the building with a rich and brilliant decoration.* They were frequently used in all the early buildings at Olympia. The pediment of the Heraeum was surmounted by a remarkable acroterium of painted terracotta, a colossal and unique ornament, almost circular in shape, set off with designs in various colours. The Treasury of Gela was decorated in the same way. It was built by the inhabitants of the Sicilian town about 582 B.C., and although constructed of limestone still preserved the primitive style of ornamentation. All the upper part of the edifice was coated with coloured mouldings of terra-cotta fastened by large bronze pins to the cornice and the frame of the pediment, and finished by a cymatium of the same material with shoots to carry off the water. "It is difficult to realise the original and attractive appearance this rich and harmonious decoration gave to the building, and what a wonderful resource for the ornamentation of wooden temples Greek architecture of the sixth century found in the use of coloured terra-cotta." †

During the whole of the archaic period the entire Greek world decorated its buildings in this way; as far as the remotest districts of Magna Graecia and Sicily, at Gela and at Selinus, at Syracuse, Metapontum and Croton, their temples were covered with coloured terracotta mouldings, and even when stone took the place

---

* Rayet and Collignon, *Histoire de la Céramique Grecque*, p. 380. (See also Prof. Middleton, art. "Templum," in Smith's *Dict. of Greek and Roman Antiquities*.—E. R. P.)

† Rayet and Collignon, pp. 382, 383.

of wood in their construction the same methods of ornamentation continued to exist, at all events, wherever the materials were of inferior quality. When, in the fifth century B.C., marble almost everywhere took the place of limestone, the reign of terra-cotta was at an end, but Greek architects still left it a certain share in decorative work. At the present time it seems to be coming into favour again, and it may not be unprofitable to notice casually the antiquity of the skilful combinations which have recently blended iron and terra-cotta so as to produce marvels of lightness and colour.

The history of ancient sculpture is equally indebted to the excavations at Olympia, which have brought to light an enormous number of relics giving, as it were, an epitomised history of plastic art in Greece from its earliest days. In the lowest layers were found thousands of small objects in clay and bronze, ex-votos accumulated at the foot of the primitive altars in the Altis by generations of worshippers, the rude technique of which is a sufficient proof of their great antiquity. There are tripods, little figures of animals, horses, stags, cows and birds, likenesses of men of the most primitive kind,—all hammered out or made of different plates of metal riveted together, and all plainly bearing the impress of Oriental art. On this western side of Hellas, as well as on the coasts of the Archipelago, we find traces of that ancient connection with Asia which lies at the root of all Hellenic civilisation. It was from Egypt and Phoenicia that the custom was derived which was so dear to the early pilgrims to Olympia,—that of consecrating their own likenesses and those of their animals at the foot of the altar, so as ever to keep before the god the memory of his devoted worshippers. It was from Asia, too, that the style of ornamentation was derived which is found on these archaic bronzes—geometrical designs, lotus-flowers and palms, strange animals, griffins and sphinxes, all borrowed from Oriental art. Yet, if the designs were derived from Asia, the greater part of the articles themselves were made in Greece, and some already display remarkable energy and life. Several were manufactured at Argos, which seems from the earliest times to have carried on continuous intercourse with the East.

Plate in Oriental style (Olympia Museum).

Among these ex-votos, the earliest of which may be dated as far back as the eighth century, a very high place must be given to a fine bronze griffin and to several metal plates in *repoussé* work, one of which is thoroughly Oriental in style and represents the Persian Artemis standing between two lions. This work undoubtedly belongs to the seventh century.

Leaving these remote times and coming down to the period of archaic art, we find figurines of bronze and clay representing the Olympian Zeus according to the type which prevailed before the time of Phidias. There is one very fine bronze head, in particular, careful in its anatomy and somewhat dry in execution, which recalls the technique of the Aeginetan masters; the work is careful almost to fastidiousness, the style archaic still, and somewhat realistic, but already full of expression. There is also a colossal head of Hera

Archaic Zeus, Olympian Bronze.

in stone, very much injured unfortunately, which is one of the earliest works of archaic art in Greece, dating from the beginning of the sixth century. In spite of the fact that it is indescribably unfinished and imperfect, we are yet conscious of a childlike effort to give it an expression of benevolence and serene majesty. We should also mention the head of a warrior, almost certainly a portrait, and the oldest piece of sculpture in marble which has been discovered at Olympia, in which many archaic details are still traceable, although the fairly delicate modelling marks the beginning of the fifth century. The most remarkable, however, among the early works are unquestionably the limestone figures which decorate the pediment of the Treasury of Megara. In this narrow space, about nineteen feet long and only two and a half feet high, the same subject afterwards so brilliantly treated by the sculptors of Pergamum—the battle between the gods and the giants—was represented by twelve figures, half the natural size; and since the limited space did not admit of sculpture in the round, the five pairs of combatants are represented in relief. Only the central pair, Zeus throwing a giant to the ground, are tolerably well preserved. The style is very archaic, the anatomy particularly undefined and conventional, and the numerous traces of red which have been found on the blue background of the tympanum show that the brush was called in to supply the deficiencies of the chisel. Still this work from the second half of the sixth century is by no means to be despised; together with the tufa figures discovered on the Acropolis of Athens it affords us the earliest example of pediment sculptures with which we are acquainted.

Before we come to the chief work of all—the most splendid discovery, perhaps, which was made in the course of the excavations at Olympia—some fine sculptures of the fourth century should be mentioned here: a charming head of Aphrodite in Parian marble, an evident replica of the Venus of Cnidus by Praxiteles, whose rounded chin, and the delicate oval of whose face, together with the slightly waving hair and sweet expression, recall the refined style of the great Athenian master; and an ideal head of an athlete found in the large gymnasium, in which the exquisite grace and perfection of finish equally point to the school of Praxiteles. Especial attention, too, should be directed to the wonderful bronze head, strikingly natural and of the boldest realism—a likeness undoubtedly of some Olympic victor, in which the artist has rendered with unusual audacity the resolute, almost insolent and brutal expression of the coarse athlete. The technique is unrivalled, and the bronze worked with the most minute care. The gymnasium which contained such masterpieces must indeed have been a marvellous museum.

The most splendid discovery at Olympia was, however, made in the ruins of the Heraeum. Pausanias, when describing the temple and enumerating the works of art which it contains, says: "There is also a Hermes in marble, carrying the infant Dionysus. The work is by Praxiteles."* It was this statue which on May 8th, 1877, was discovered among the *débris*, and by astonishing good fortune it was in an excellent state of preservation. The god, a youthful figure of the utmost grace, stands carelessly leaning against the trunk of a tree; the weight of the body falls on the right leg, and it is slightly curved so as to give a graceful, pliant undulation to the lines, and to turn to account the curve of the hip and attachment of the thigh. On his left arm sits the infant Dionysus, one of his little hands pressed against the shoulder of the

---

* Pausanias, v. 17. 3.

god, while the other is raised with a gesture of eager desire, and all his little body trembles with impatience. With the right hand and arm, now unfortunately lost, Hermes held up to the child some object which excited his desire—most likely a bunch of grapes; at least, such seems to be the conclusion we should draw from certain copies of this group, so famous even in antiquity, which are works of little value in themselves, but are apparently inspired by some faint memory of Praxiteles.

It is almost needless to praise this exquisite figure, so perfect in its execution, and so in-.. comparable in its grace; nothing is more piquant than the contrast between the almost feverish excitement of the child and the quiet attitude of the god, slightly bending his half-smiling face. It

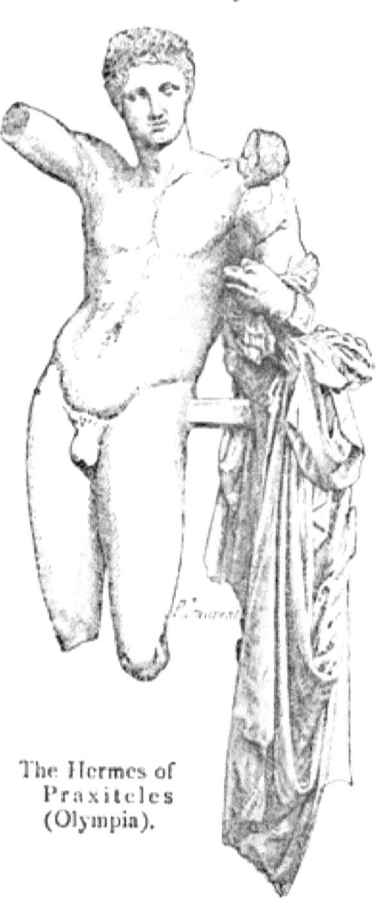

The Hermes of Praxiteles (Olympia).

is impossible, however, to dwell too strongly on the importance which the possession of an original work by one of the great Greek sculptors has for the history of art—a work, too, by one whom the unanimous verdict of the ancients placed by the side of Phidias. Before this discovery was made we only caught a glimpse of the art of Praxiteles through inadequate copies of his works, and although we might gain from these copies some notion of the Venus of Cnidus or of the Apollo Sauroctonus, and although the Satyr of the Louvre, dug up on the Palatine, seemed to some a fragment of the master's own work, yet these were but poor relics of one of the most famous sculptors of antiquity. Now we possess a revelation of his genius in an authentic work, and if we remember with how few artists this is the case (Lysippus, Polycleitus, and others are not known to us by any undoubtedly original work), we shall realise the importance of this Olympian statue. It is true that the Hermes is not one of the most famous works of Praxiteles, it did not enjoy in ancient times the renown of his Aphrodite, his Eros, or his Satyr; what then must those masterpieces have been, if this second-rate statue seems to us so exquisite?

We cannot here undertake an analysis of the characteristic features of the genius of Praxiteles, but nevertheless we must briefly state the new elements which he introduced into the conception of Hellenic art. The ancients, whose artistic criticisms are always couched in rather vague terms, say that he excelled in fidelity to nature without, however, falling into realism, and that he permeated his works in marble with the moods of the soul.* The deities he represents have, in fact, nothing in common with the gods of Phidias, whose sovereign majesty rises to such a height above

* Diod. Siculus, *Frag.*, lxxvi.

all mortal woes; the fourth century, in which Praxiteles
lived, no longer possessed the reverence and the faith
which could inspire such lofty conceptions. The increase
of luxury, the ease and comfort of life, the refinements
of intellectual culture, had deadened the passions and
quenched the enthusiasm of the century before. To
this new world the gods of Olympus were no longer
superhuman beings, only to be approached with religious
awe; the chisel of Praxiteles is quite at home among
them—he brings them down to the proportions of
mortal men, and animates them with all the passions
and the feelings of human beings. Apollo becomes
a youth playing with a lizard, and Venus a beautiful
woman who is letting her drapery fall as she enters
her bath. What the sculptor demands of them is not so
much that they should express the noblest conceptions
as that they should reveal the most perfect grace.
All these figures possess a harmonious grace and a
youthful pliancy which is almost effeminate. " In
order to draw attention to the undulating and flexible
lines of the body, the figure is thrown out of the per-
pendicular, its weight is cast on one foot, and it leans
effeminately against the trunk of a tree. In this way
the master breaks the vertical lines of the body and legs,
and diversifies the symmetrical aspect of the limbs,
while he calls attention to the hips and to the elegance
of the modelling in general."* The Hermes, the Apollo
and the Satyr are all represented in this quiet and
careless attitude; it is characteristic of the works of
Praxiteles.

It is not less interesting to note the attention paid to

* Rayet, *Mon. de l'Art Antique*, notice de la pl. 45.

detail. As a rule the subjects the master chooses for his chisel are in early youth, but none of the athletes whom the sculptor might take as his model could offer the harmonious and pleasing contours in which he delighted—the Apoxyomenos of Lysippus, so well

Head of the Hermes of Praxiteles.

built, so graceful in his strength, looks a rustic beside the Ephebi of Praxiteles. There is more softness and bloom about the flesh of the great Athenian sculptor, the play of the muscles can scarcely be detected under the delicate skin. It was from his female models that Praxiteles drew these rounded limbs, this delicacy and grace, and he has blended the results of these various studies so as to create an indefinable charm.

The part of Praxiteles' work which was most admired was the head. It was there that he displayed that power of expressing feeling which no sculptor of antiquity possessed to the same extent. The head of the Hermes certainly does not belie this reputation, while at the same time it shows plainly that the group is a work of the master's youth, and bears marks of the influence of Myron. Still, in spite of some defects and of a certain dryness in the execution, the workmanship here is wonderful, the artist's skill is so perfect that all trace of effort has disappeared, and the means employed are so simple that the fruit of prolonged thought and of patient study seems to have been thrown off carelessly and to have cost no pains.

If the excavations at Olympia had given us this one statue only, Germany might be justly proud of having undertaken them, but they are also interesting from quite another point of view. Artistically they may be reckoned among the most important discoveries of recent years, while archæologically considered, they are encouraging as showing what treasures are still hidden beneath the soil of these regions, and what advantages science and art may still gain from investigations which seem to some useless and costly. We can only hope that the spoils the French may win from Delphi will be in no respect inferior to those which have made the excavations undertaken by the Germans at Olympia for ever memorable.

CHAPTER VIII.

THE EXCAVATIONS AT ELEUSIS (1882-89).

REFERENCE BOOKS FOR ELEUSIS:—

 Philios, "Inscriptions d'Eleusis" (*Ephemeris Archaiologike*, 1883—1888).
  " "Monuments d'Eleusis" (*Eph. Arch.*, 1883—86, 1888).
  " "Rapports sur les Fouilles d'Eleusis" (*Praktika*, 1882—1888).
 Tsountas, "Inscriptions d'Eleusis" (*Eph. Arch.*, 1883, 1884).
 Blavette, "Fouilles d'Eleusis" (*Bull. de Corr. Hell.*, 1884-5).
 Foucart, "Le Culte de Pluton dans la Religion Eleusinienne" (*Bull. de Corr. Hell.*, 1883).
  " "Inscriptions from Eleusis," 1880, 1884, 1889.
 Decharme, *Mythologie de la Grèce Antique*. Paris, 1886.
 Benndorf, *Anzeiger der Phil. Hist., Classe d. Academie in Wien*, Nov. 1887.
 Furtwängler, *Arch. Gesellschaft in Berlin*, July 1887.
 Reinach, *Gazette des Beaux-Arts*, 1888.
  " *Antike Denkmäler*, pl. 34.
 Louis Dyer, *The Gods in Greece*.
 Prof. Gardner, "New Chapters in Greek History" (*Eleusis and the Mysteries*).

AMONGST the excursions which the tourists who visit Athens are in the habit of making, there is none easier or more enjoyable than that to Eleusis. On leaving the city the road, which follows nearly the same course as the ancient Sacred Way once trodden by the solemn procession of the initiated, passes through the great grove of olive-trees whose pale foliage stretches in a dusky band across the thirsty plain of Attica, from Cephissia down to the sea. Then, entering the mountains, it crosses the outlying spurs of Parnes by the charming pass of Corydallos, where the little monastery of Daphni lies with its graceful Byzantine church, and as it issues from this pass, suddenly there bursts upon the traveller's sight the dazzling surface of the sea, the blue waves washing Salamis, and the little village of Lefsina, the ancient Eleusis, lying at the farther end of its tranquil bay, while in the distance there rise the hills of Megara and the lofty peaks of Cithaeron. Passing by the side of the salt lakes once sacred to Demeter and Cora, the road runs along the coast until it reaches the village of Lefsina, now exclusively inhabited by Albanians in their picturesque costume, who, short as the distance is from Athens, scarcely understand a word of Greek. Here the famous temple of Demeter once stood, and here ancient mythology localised the legend of the goddess, one of the most beautiful and touching ever conceived by the imagination of the Greeks.

The story of this divine tragedy, at once epic and mystical, in which the Greeks symbolised the changes undergone by vegetation, and the life and death of nature, should be read in the beautiful hymn to Demeter discovered about a hundred years ago in a library at Moscow. When Pluto, so the poet's story runs, carried off the maiden Cora as she was plucking flowers in the sunny fields of Nysa, her piercing shrieks reached the heights of Olympus, and Demeter, her mother, shuddering as she heard that imploring cry, darted to earth in search of her beloved child. Then began the long and grievous passion of the afflicted mother. Speechless and inconsolable, she wandered over earth and heaven,

The Demeter of Cnidus (British Museum).

seeking everywhere for some intelligence of the child

she had lost. "For nine days," says the poet, "the dread goddess wandered over all the earth, holding flaming torches in her hands; in her grief she tasted neither ambrosia nor sweet nectar, nor did she give her body to the bath." When Helios, the sun who looks on all things, told her at last that Pluto was her daughter's ravisher, and that Zeus himself had consented to the deed, the angry goddess left Olympus for ever. Casting aside her immortal beauty, she concealed the bitter grief which was gnawing at her heart beneath the form and vesture of an aged woman, and wandered sadly through the fields and towns of men, consumed by longing for the child she had lost, until at last she came in her sorrow to ask for shelter from the mythical king of the little village of Eleusis.

Exhausted by grief and bowed down with weariness, the goddess seated herself by the well, in the shade of an olive-tree. Here, in a charming scene which recalls the most delightful customs of Oriental hospitality, the daughters of the house, coming to draw water at the well, find the old woman, question her, and take her back with them to their father's dwelling, where she is entrusted with the care of the king's infant son. Still nothing can relieve the goddess' profound sorrow, neither the kindness of her hosts nor their gentle words. "Seated there, with her hand she held her veil before her face, and speechless in her grief, remained in her place, greeting none either by word or gesture, but without a smile, without touching meat or drink, she sat consumed by longing for her deep-girdled child." The divine glory of Demeter, however, could not long remain concealed: she betrayed herself, almost against her will, and the humble handmaiden, suddenly transfigured, stood revealed to her hosts as the all-powerful goddess who bestows fruitfulness upon the earth and gives corn to men, and promised that in the temple which was to be erected to her on the hill of Eleusis she would herself teach her holy mysteries to the mortals who had given her a kindly welcome.

Still this is but an episode in the story, though no doubt the most cherished at Eleusis, for the divine passion of this Mater Dolorosa of ancient times was not yet at an end. The goddess would not return to Olympus, and her anger had disastrous consequences: the earth became unfruitful, the seed sown did not spring up, for the curse of Demeter was upon the world; and the injured mother would not be reconciled until her beloved child was given back to her. By the command of Zeus Cora returned to the light of day, but alas! before she left the lower world she had secretly eaten the pomegranate seed, the sacred pledge and symbol of her union with her dread lord, and henceforth she could no longer pass her whole life with her mother in Olympus. For one-third of the year she must again go down beneath the earth; but, when "in the sweet spring-time the earth is clad with a thousand kinds of flowers," each year the young goddess was to leave that subterranean darkness and ascend once more to the divine abodes. At the sight of her daughter Demeter was appeased, the world resumed its regular course, the earth was covered with leaves and flowers, and joy sprang up all around. Faithful to her word, before she returned to Olympus the goddess instructed the men of Eleusis in her divine knowledge, and taught

Triptolemus, her first priest and the founder of her worship, the marvellous secrets of agriculture which he was to spread throughout the world; above all she initiated him into those august mysteries, "which we may neither neglect, nor penetrate, nor divulge, for reverence for the gods restrains our voice.'

It is hardly necessary to explain the meaning of this beautiful legend. It arose from the sight of the phenomena of vegetable life, from the mourning of nature during the winter, followed by the new birth of spring. This sight, in which our colder reason only seeks for natural causes, was lit up with the most marvellous colours by the wonderful imagination of the Greeks. For them "nature was full of passionate and living energies, of divine forms sensible of joy and of sorrow, and the different phases of vegetable life became in their eyes the wondrous acts of a drama which was at once divine and human." * The annual descent of Cora into the lower world and her return to the light symbolised the seed falling upon the earth and disappearing into its bosom, and then coming to life again and flourishing in spring; the grief of Demeter, bereft of her child, represented the desolation and barrenness of the earth during the winter. To this assemblage of physical notions were soon added moral ideas of a loftier kind: the life and death of man, and even the problems of the destiny of mankind, were linked with the vicissitudes of the earth, with the life and death of nature; soon the thought of the future life prevailed over the touching story of the divine tragedy, and the combination gave birth to the mys-

* Decharme, *Mythologie de la Grèce Antique*, p. 387.

teries of Eleusis. The author of the hymn to Demeter, himself undoubtedly one of the initiated, makes us feel the lofty aim of these mysterious rites when he closes his tale with the remarkable words, which bear the marks of such profound piety: "Happy among the dwellers upon earth is he who hath beheld these mysteries, but he who is uninitiated and hath no part in these rites hath never the like destiny, even when death hath made him descend into the gloom of the lower world."

I.

The little town of Eleusis, where Demeter lived and suffered and manifested her divine splendour, where she taught men the secrets of her worship, was a veritable Holy Land to the Greeks. Everything there recalled the memory of the goddess and bore the mark of her footsteps. "Here was shown the well by which she sat and the stone on which she rested; there, the house which sheltered her in her distress and the site of the first temple she commanded to be built. All around the town, the rocks and the slightest irregularities in the ground bore names which recalled the sacred legend to pious souls." On one side stood the wild fig-tree near which Pluto, carrying off Cora, had gone down to his subterranean kingdom, and the fountain Callichoros round which the women of Eleusis had for the first time danced in chorus to the honour of the goddess, on the other the field of Raros the father of Triptolemus, the first land ever sown; here was the threshing-floor of Triptolemus himself, with the altar consecrated to the hero, and there the tomb of Eumolpus, one of the earliest priest-kings of Eleusis, and ancestor of the great sacerdotal family which administered the mysteries of the temple ; lastly, yonder was the *temenos* of the goddess with the shrines it contained. All Greece came each year to these venerable monuments, in order to testify its gratitude to the goddesses who had given corn to mankind and through many centuries the Greek cities, in obedience to the strict commands of the Delphic oracle, consecrated the firstfruits of their harvests at Eleusis. Athens especially, and her allies, never neglected to make these pious offerings, and even in the time of Hadrian the Panhellenic assembly, representing the entire Greek world, in conformity with the old regulation of the fifth century did homage to the benevolent goddess who had taught men the arts of agriculture, and to Athens, who by sharing this gift with the other Greek tribes had deserved to be commended by the god of Delphi to the gratitude of the Hellenes as the μητρόπολις τῶν καρπῶν.

Athens, indeed, had early adopted the worship of Demeter, and devoted itself to the administration and embellishment of her sanctuary. From the sixth century B.C. a large building for the celebration of the mysteries had been erected in the *temenos*, by the side of the primitive temple built on the hill, and also some early buildings, of which traces still remain in the plain. Recent excavations have recovered the remains of this rectangular hall, which was preceded by a vestibule and supported on twenty-five columns, afterwards buried under the ruins of later buildings, as well as the walls of unbaked brick of the rather narrow *temenos*, which in the time of Pisistratus inclosed the shrines of Demeter. In later days, at the time when under the rule of Pericles Athens was being filled with magnificent buildings,

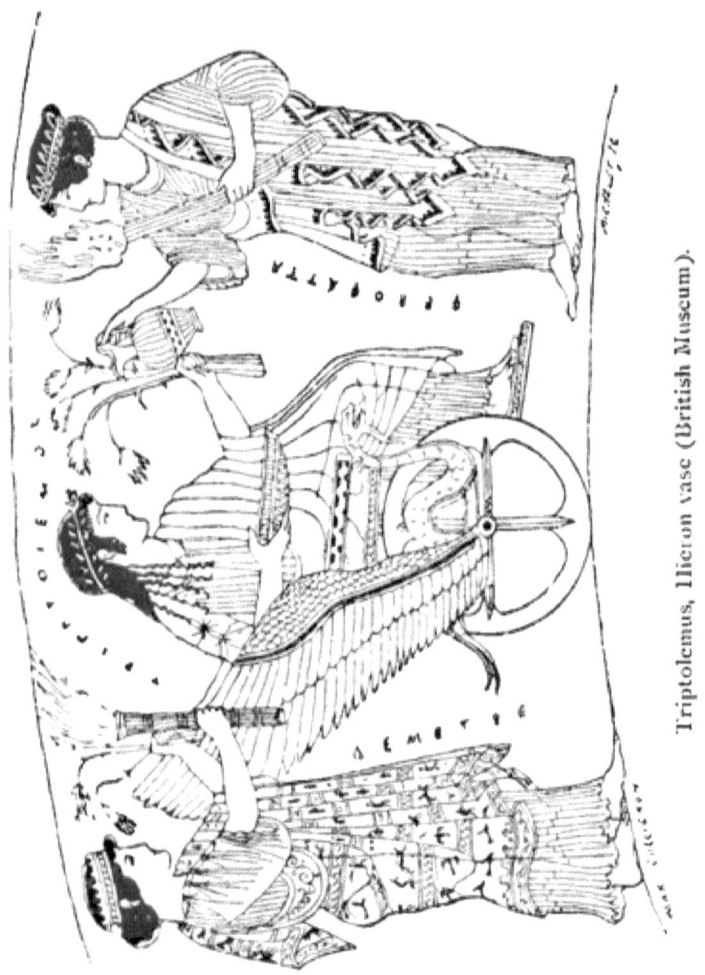

Triptolemus, Hieron vase (British Museum).

Eleusis also had a share in the splendid works which owed their rise to the great orator. The primitive temple, built on an isolated hill commanding the plain, had suffered so severely at the hands of the Persains

that it had been necessary to rebuild it entirely, but the
Hall of Initiation, which had been destroyed by the same
invaders, had only undergone temporary repairs. In its
place, in the sacred enclosure, which had been enlarged
and fortified with strong towers, Pericles had an
immense hall built for the celebration of the mysteries,
after the plans of Ictinus, the architect of the Parthenon,
three of whose pupils superintended the work. In the
fourth century Athens seems still to have taken an
active interest in the sanctuary of Eleusis, and the great
orator Lycurgus, who directed its finance in the second
half of the century, seems to have been especially
devoted to the worship of Demeter. A member of an
old priestly family, thoroughly acquainted with all
religious matters and eager to restore the ancient
ceremonies, he instituted important works within the
sacred enclosure. The inscriptions of this period, and
the accounts of sums expended on the buildings, recall
to us the existence of the numerous edifices which were
then crowded together within the *temenos*. There was
a temple of Pluto, built at this very time; there were
the treasuries of the two goddesses, the dwellings of
the priests or Neocorion, and that of the priestly family
of the Ceryces, as well as the houses of the priestess
and the daduchus; there were altars set up at various
points within the enclosure, the bouleuterium or council-
chamber, and lastly the Propylaea, dating from the
middle of the fourth century, which formed a stately
entrance into the newly repaired and fortified *temenos*.
It was already in contemplation to erect a Doric portico
before the hall of the mysteries, and this building was
completed towards the end of the century, under the

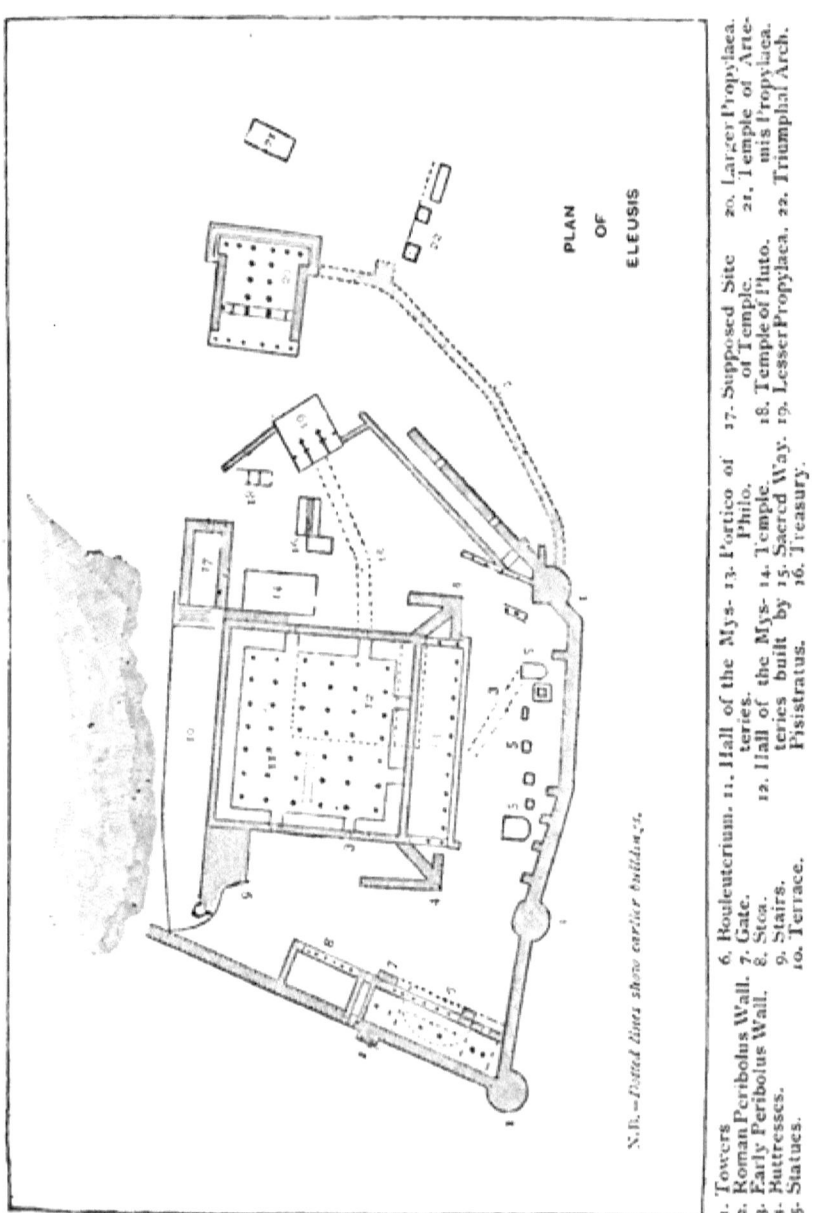

N.B.—*Dotted lines show earlier buildings.*

1. Towers.
2. Roman Peribolus Wall.
3. Early Peribolus Wall.
4. Buttresses.
5. Statues.
6. Bouleuterium.
7. Gate.
8. Stoa.
9. Stairs.
10. Terrace.
11. Hall of the Mysteries.
12. Hall of the Mysteries built by Pisistratus.
13. Portico of Philo.
14. Temple.
15. Sacred Way.
16. Treasury.
17. Supposed Site of Temple.
18. Temple of Pluto.
19. Lesser Propylaea.
20. Larger Propylaea.
21. Temple of Artemis Propylaea.
22. Triumphal Arch.

PLAN OF ELEUSIS

administration of Demetrius of Phalerum, by the architect Philo of Eleusis. At the same time all the space between the new portico and the *temenos*-wall was filled up so as to form a vast terrace, where in the marble-paved court there rose the statues and monuments which the grateful inhabitants of Eleusis had set up in honour of the benefactors of the temple.

The Roman period saw the prosperity of Eleusis increase still further. Appius Claudius Pulcher, the friend of Cicero, erected the lesser Propylaea, and Munatius Plancus won for himself statues from the grateful Eleusinians, possibly because he had the portico built which ran along the southern wall of the *temenos*. The reign of Hadrian was, however, the time when the buildings increased most rapidly. This emperor, learned, artistic, and literary in his tastes, was passionately attached to Greece, took a keen interest in all the relics of its glorious past, and gladly restored everything which had formerly existed there. He heaped benefits upon the Athenians and filled their city with buildings, until they believed the golden age of their country had returned as they saw this master of the world wearing the Greek dress, speaking the Greek tongue, and presiding in the archon's robes over the festival of Dionysus. Enamoured, too, of all that was strange or foreign, he wished to be initiated into the mysteries, and one of his teachers in philosophy was a member of the priestly nobility of Eleusis. Marcus Aurelius, Lucius Verus and Commodus, were also initiated, and the temple was naturally the object of imperial liberality. The sacred enclosure was enlarged for the last time, and an imposing entrance made; the

Great Propylaea were built in imitation of those on the Acropolis of Athens, while the hall of the mysteries was decorated afresh, and on the terrace in front of the portico of Philo, statues of priests and priestesses, and monuments of every kind set up in honour of Roman benefactors, occupied the place of honour in the enclosure. In the neighbourhood of the sanctuary other buildings were erected—here a graceful villa, where fine frescoes have been discovered, and there, close to the gateway, a temple of Artemis Propylaea. Further on were the baths and on one side a triumphal arch, consecrated by the Panhellenes to the goddesses and the Emperor Hadrian.

Unhappily, in one point the good fortune which marked the explorations at Olympia was wanting at Eleusis, for there was no guide-book to describe the details of the famous shrine and to aid in the identification of so many buildings of widely different date. Pausanias had undoubtedly seen them, for he had been initiated, and consequently was able to enter the sacred precincts and to be present at the ceremonies there. In spite of the prohibition laid on all who took part in the mysteries, which forbade them to divulge any of their secrets, he seems to have resolved to describe the temples of Eleusis rather than leave a blank in his book. A seasonable dream, however, reminded him of his vow, and of the formal prohibition which not only forbade the uninitiated to see anything of the sacred buildings, but even to become acquainted with their arrangements from description. The other travellers of ancient times are not more communicative than Pausanias. Strabo, who does not seem to have been initiated, only described the buildings which he could see over the *temenos*-wall as he passed—that is, the temple and the great hall of the mysteries, which was so large that it could hold as many spectators as a theatre. Excavations alone could be expected to raise to some extent the veil which had so long hidden the sanctuary of Demeter. In 1815 the Society of Dilettanti which was formed in London for the study of the existing remains of ancient Greece, and which has rendered valuable services to archæology, undertook the exploration of the ruins of Eleusis, and made a first statement, necessarily imperfect and to some extent inaccurate, of the buildings which still existed under the houses of the village of Lefsina. In 1860-61 M. Lenormant sank some shafts near the portico of Philo, laid bare a part of the temenos and explored both Propylaea. It was not, however, until 1882 that the Athenian Archaeological Society undertook methodical explorations at Eleusis. The Greek government expropriated the occupants of the modern houses built above the ruins, and when these were pulled down it was for the first time possible to clear out the *temenos* completely, and thoroughly to explore the sacred precincts of Demeter. The excavations were actively carried on from 1882 to 1889 under the direction of M. Philios; and if the cost was considerable—the Greek government paid 100,000 francs as compensation and the Society 90,000 francs for the expenses of the work itself—important results have been obtained. Interesting works of art have been discovered, as well as numerous and important inscriptions, which help to throw a light on the history of the temple and the details of its administration; more than this, the excavations have made it possible to ascertain the plan and arrangement of the buildings within the *temenos*, and although much still remains uncertain, we can now form a fairly accurate conception of the great festival celebrated at Eleusis in honour of Demeter, of the priesthood which presided over her worship, and of the principal buildings connected with the cultus of the great goddesses.

II.

When, after the victory of King Erechtheus over the Eumolpidae, Eleusis entered the Attic Confederation, important privileges were secured to the inhabitants of the little town, and they preserved a certain independence in all that concerned their religious rites and the administration of the sanctuary. The two great priestly families, the Eumolpidae and the Ceryces, remained in possession of all that appertained to the mysteries and initiatory rites, they retained the right of collective action under certain specified circumstances, and their official representatives, the hierophant, chosen from among the Eumolpidae only, and the daduchus, always appointed from among the Ceryces, were the two heads of the Eleusinian priesthood. These two important officials, who were appointed for life, jointly determined all matters concerning the cultus of the two goddesses; in later times the sacred herald, also chosen from among the Ceryces, took the third place in this sacerdotal hierarchy, and the same family also furnished a fourth dignitary, the priest of the altar (ὁ ἐπὶ τῷ βωμῷ). Parallel to the priestly hierarchy, several priestesses played a part in the cultus of Eleusis; in the inscriptions we find the hierophantis mentioned, and also the priestess of Demeter, the representative of a more ancient cultus, who figures as eponym on several monuments, and whose privileges corresponded to those of the hierophant. The most important part of the religious administration was in the hands of the great family of the Eumolpidae, whose members possessed the right of interpreting the sacred laws of Eleusis to the exclusion of the Ceryces ; as the guardians of the sacred books and of ancient tradition, they could form themselves into a tribunal to try cases of impiety ; they alone regulated the sacrifices, which were their property ; and the Athenian law itself solemnly reserved to them the right of administering the worship of the sanctuary, and of taking all necessary measures, in conformity with their institutions.

Athens, notwithstanding, had early laid her hand upon the sanctuary. One of the ten strategi, the στρατηγὸς ἐπ' Ἐλευσῖνος, was put in especial command of this district, in order to protect the temples and to defend this frontier, the most exposed to attack ; and although the town had no permanent garrison, every year the ephebi were encamped there to complete their military training in the fortress. In the religious worship of Eleusis, over against the hereditary priesthood the King-Archon represented the city. He offered solemn sacrifices in the name of the State ; with the assistance of the four epimeletae (superintendents) of the mysteries, two of whom were appointed from among the citizens of Athens, he watched over the procession of the initiated, and after the festival he presented a report to the Council upon the celebration of the mysteries, noting any infractions of its rules which had been committed. Besides this, ten members of the Council of the Five Hundred were responsible for the due performance of those among the ceremonies of Eleusis in which the city was concerned. The administration of the sacred treasure was, however, the department in which the State reserved for itself a preponderant influence. The registers of accounts which have come down to us in inscriptions, and the inventories of the treasure, show in what the revenues of the sanctuary consisted, and what use was made of them. The receipts include the rents of the lands belonging to the temple, the proceeds of the firstfruits consecrated to the great goddesses, and the offerings laid up in the treasuries of Demeter and of Cora. The disbursements, which were numerous, were usually on account of the building or repairing so often undertaken in the sanctuary, the expenses of the festivals and the ordinary rites, the maintenance of the temple staff, who according to ancestral custom, received a part of their pay in kind, and the prizes given to the victors in the games. Almost all this part of the management was under the control of Athens. Hieropoioi, appointed from among the Athenians for one year, were responsible for the temple and for all the outlay connected with its ritual; seven epistatae of Eleusis and two treasurers of the goddesses, who remained in office for a period of four years, were entrusted with the superintendence of the works and the necessary payments; these, too, were chosen from among the Athenians. Lastly the King-Archon in conjunction with the epistatae and epimeletae, was charged with the letting of the land belonging to the temple of Eleusis, and the Athenian people reserved the right of control over all these magistrates and the power of fining them if they were unfaithful to their trust. As to the festivals celebrated in honour of Demeter, they were divided between Athens and Eleusis. At the foot of the Acropolis a sanctuary, the Eleusinium, was consecrated to the great goddesses, and it was beneath its colonnades that the celebration of the mysteries began. Other assemblies took place at Eleusis itself; by the side of the old national festival (πάτριος ἀγών), which the Eleusinians had been careful to keep up, other solemnities brought all the Athenian citizens together at intervals of two or four years respectively. There were gymnastic and musical contests, and theatrical representations; in the management of these, too, the Athenian people interfered by their decrees, and the orator Lycurgus in particular introduced some important innovations. It was, however, to the mysteries, above all, that Eleusis owed its fame.

By the side of the public rites, there were secret ceremonies in the Greek religion from which all were excluded who had not satisfied certain specified conditions, and which were called mysteries on account of the secrecy imposed upon the initiated. As a matter of fact, there is no religion without mysteries; religious feeling is always eagerly desirous of penetrating the unknown, and by the side of the piety of habit and custom there is always to be found a more enthusiastic and more mystical faith which aspires to penetrate the dark enigma of human destiny. It was naturally from the worship of the chthonic divinities who presided over the perpetual succession of life and death in nature that the Greeks demanded the secret of the life and death of men ; and it was this that Hellenic piety sought at Eleusis. Under the influence of certain foreign cults, and of Oriental mysticism in particular, under the influence of the Orphic school, which early associated the more exciting cultus of Dionysus with the adoration of Demeter and Cora and imposed upon the initiated the severe practices of asceticism, under the influence too of philosophy, which introduced into them some of its doctrines, the mysteries of Eleusis acquired a much loftier meaning and a far wider range than the simple homage which was rendered to the great goddesses in early times. "They seem to have formed," says M. Renan, "the really serious part of the religion of ancient times ; and they exercised a strong attraction and considerable moral influence over the pious souls of those days."

Unfortunately our acquaintance with these mysteries is very imperfect. The vow of secrecy imposed upon the initiated was scrupulously observed ; and a few passing indiscretions, a few chance allusions, a few parodies springing from the imagination of Aristophanes, which we must be careful not to take too literally (for the example of Alcibiades shows that the mysteries could not be mocked at with impunity), would furnish us with very scanty information, if the Fathers of the Church, in their fiery assaults upon paganism, had not, fortunately for us, often attacked the mysteries of Eleusis. It is true that in their somewhat intemperate zeal Origen and Clement of Alexandria forgot that the sanctuary of Demeter sheltered all that was noblest in

Greek religious belief, and welcomed without criticism more than one foolish tale, and misinterpreted more than one mysterious symbol; still we owe to them a certain number of curious details which enable us to give a fairly accurate account of what the mysteries really were.

We can well imagine that the divine revelation which the worshippers sought at Eleusis could not suddenly burst upon their eyes and ears, but must be reached gradually, through the successive steps of a slow initiation. Before they could be worthy of understanding the highest mysteries, and capable at the same time of interpreting the obscure symbols which accompanied the full possession of the secrets of the gods, they must have passed through a preliminary course of instruction and have received a preparation which was at once moral and dogmatic. This was imparted in the lesser mysteries, also called the Mysteries of Agrae, from the name of the hill on the banks of the Ilissus on which they were celebrated, at the beginning of spring. They included purification, symbolical of the moral purity exacted of those who desired to become initiated, possibly also a kind of solemn confession, such as took place in the mysteries of Samothrace, and above all prayers, litanies in honour of the gods, and the recital of the sacred legends—a kind of elementary instruction which was designed to make the initiated acquainted with the mystic story which would afterwards be enacted before their eyes.

Seven months later, in the autumn, complete initiation was received at the greater mysteries, and from that time the worshipper belonged to the holy brotherhood of the Mystae. Nevertheless he had not yet attained the highest grade of initiation; for this a new ceremony, the ἐποπτεία, was necessary, and this could only take place after an interval of a year. Then, however, he had reached the goal of his pilgrimage: the ἐποπτεια was the final and divine revelation which by its virtue led to complete religious perfection, on which account it was also named τελετή.*

Naturally there could be no question, at the admission of the initiated, of submitting the considerable numbers who presented themselves to any severe examination. In the solemn proclamation made by the hierophant at the beginning of the greater mysteries, only those were excluded who had committed a murder, or were suspected of impiety or magic, or had been born among barbarians. Each of the candidates, however, must have a kind of godfather, chosen from among the initiated, called the mystagogus, whose duty it was to explain to the neophyte the symbols he would see and the sacramental words which accompanied them. The heads of the Eleusinian priesthood reserved to themselves the right of acting as mystagogi for the greater number of candidates, and in this way they could exercise a kind of control and render the ceremony of initiation more solemn and severe; but the ordinary mystagogus had fewer scruples, and more than once slaves and courtesans gained access to the mysteries.

At first the greater mysteries were celebrated every five years, afterwards they took place annually in the

---

* Professor Gardner identifies τελετή with μύησις—the initiation properly so called.—*New Chapters in Greek History*, p. 586.—E. R. P.

month Boedromion, and the Athenian government attached so much importance to their regular celebration that in time of war a sacred truce was concluded for the protection of the festival. It lasted no less than fourteen days, and took place partly at Athens and partly at Eleusis. On the 15th of the month the mystae assembled under the colonnades of the Eleusinium at Athens to listen to the proclamation of the hierophant, enjoining purity of body and mind on all who wished to take part in the festival; the next day they went together to the sea-shore to undergo the prescribed purifications and ablutions; the three following days were spent in sacrifices and expiatory ceremonies, by which the mystae were at last made worthy of the great mystery which was about to be revealed to them. Finally, on the 20th, the sacred procession left the city and marched to Eleusis along the Sacred Way, bearing with it in great pomp the statue of Iacchus, the divine child whose worship was introduced into the Eleusinian cultus in connection with the Orphic doctrines, while, to the music of the flute and of the sacred hymns, from thousands of enthusiastic voices there rose the name of the youthful deity, the intermediary between the great goddesses and their fervent worshippers.

The procession left the Eleusinium in the morning, and slowly made its way towards Eleusis along the Sacred Way, bordered with tombs and altars, with chapels and shrines recalling the great memories of the sacred story; renewed purifications took place at the salt lakes, and as night fell the procession entered the holy city by the light of torches. By a contrast which was not rare in ancient times, a comic element entered into this solemn ceremony. At the bridge over the Cephissus a veritable masquerade awaited the procession of mystae; the peasants from the neighbouring country, with masks on their faces, saluted them as they passed with a running fire of jests and banter and scurrility to which the initiated did not fail to respond; it was a regular carnival scene, which satisfied the thirst for clamour and merriment which sometimes bursts out in the midst of the most austere religion.

After the arrival of the procession at Eleusis the initiatory ceremonies and the mysteries themselves began under the direction of the priests. There were purifications and expiatory rites designed to break down the barrier between men and the gods, sacrifices and processions accompanied by hymns and dances, in which the excitement of the worshippers sometimes rose to frenzy; above all there were nocturnal festivals to strike the imagination of the faithful, sacred and mystic vigils, and spectacles of extraordinary magnificence which were intended to excite profound emotion in the minds of the worshippers. The principal scenes in the myth of Demeter, her sojourn on earth and her passion, formed the chief subject of these symbolical representations, in which the Mystae were at once actors and spectators. To these were added sacred stories, mysterious and enigmatical sentences, intelligible only to the initiated, which were all connected with the mystic tale. All the details of the legend gave occasion for a picturesque symbolism which acted powerfully upon the imagination. The worshippers imitated the action of the goddess, and roused within themselves the feelings of joy and sorrow which she had felt; for nine days they fasted like her, and during this time they underwent purifications, they wandered, torch in hand, as she had done when searching for her daughter, and rested at the spots consecrated by the sacred story; then on the tenth day the fast was broken, and in imitation of the goddess they drank the *cyceon*, a drink composed of meal and water and pennyroyal, and joy took the place of the gloom and sorrow of the days before.

The mysterious rites which followed within the sanctuary were of two different kinds; commemorative acts performed by the priests and the initiated, and certain representations of a dramatic character acted before them. Among the first, the most important was a sacred meal which the mystae took in common—a real mystical communion in memory of Demeter, and among the second, the mystery or miracle-play which represented the legend of the great goddesses. The story of the poets became a sacred reality at Eleusis, "rendered palpable and visible by the different acts of what Clement of Alexandria calls a mystic drama."* As in the mysteries of the Middle Ages, the different scenes of the sacred story were represented in a series of wonderful pictures, and the priests in magnificent robes were the principal actors in this silent and imposing tragedy. The illusion was heightened by certain stage-effects produced by theatrical machinery, and strange formulae and mysterious words formed an obscure comment on the symbols presented to the eyes of the initiated. At one time the voice of the hiero-

---

* Decharme, *Mythologie de la Grèce Antique*, p. 396.

phant would be heard from the depths of the hall, and in the glow of the torches the high priest proclaimed, "The goddess has brought into the world the divine child, the mighty has given birth to the mighty." At another, the epoptae were shown, in the words of Origen, "as the great, the wonderful, the most perfect object of mystical contemplation, an ear of wheat reaped in silence." Then the silence was broken by the piercing cries of Demeter calling her daughter, answered from the depths of the hall by the dull sound of the brazen instruments. On the return of Cora there all was changed to gladness, and then followed choric dances, orgiastic ceremonies and frenzied delight.

It is easy to grasp the general meaning of these symbols; the life of nature and its relations with the life and destiny of man formed their ground work, but it may well be believed that these abstract ideas and this mysterious meaning were hidden from the greater number of the initiated. That which had such a powerful charm for them, and left such a deep impression on their minds, was rather the definite form in which this vague symbolism found expression. For these mystics, among whom were many women, the metaphysical meaning of the drama was quite a secondary thing. "What led them," says M. Renan in a beautiful passage, "to hasten in throngs to weep for Adonis? The desire of bewailing a young god who had blossomed too soon, of gazing upon him stretched on his funeral couch, cut down in his prime, his head drooping languidly, surrounded by orange trees and plants of rapid growth which they might see spring up and die, the desire of burying him with their own hands, of cutting their hair to lay upon his tomb, of lamenting and rejoicing in turn—the desire, in short, of tasting all those impressions of transitory joys and mournful vicissitudes grouped around the myth of Adonis." *

The same kind of interest was found in the miracle-play of Eleusis, when the mystic tragedy represented Cora's stay in Hades; the initiated, too, were plunged in darkness, they too wandered for a long time in the midst of this dreadful gloom, "through terrible passages, along an endless way," in which their excited imagination saw the windings of the lower world. Lightning-flashes pierced the gloom, terrible voices mingled with fearful apparitions. "Before they reach the end," says Plutarch,† "the terror reaches its height: they shudder and tremble with fear; a cold sweat chills them." Then all at once the scene shifted, and a marvellous light fell upon the mystae. They were received into the abodes of bliss, wonderful music sounded in their ears, and divine apparitions appeared before their astonished and delighted eyes. "The veil fell," says an ancient writer, "and the deity appeared to the initiated, radiant with celestial light." We can imagine the powerful effect produced by this contrast, which, when applied to human life, represented the transition from the horrors of Tartarus to the bliss of the Elysian Fields.

We are told, too, of discourses which were delivered during the mysteries; and it may be asked whether these were sermons like those of Christian preachers, conveying dogmatic instruction and intended to explain

---

* Renan, *Études d'Histoire religieuse*, p. 55.
† Plutarch, *De Anima Fr.* vi. ex p. 270.

the symbols and to reveal to the initiated lofty moral and religious truths. This is not at all likely. According to Aristotle,* the initiated learned nothing precisely in the mysteries; "they received impressions, they were put into a certain frame of mind. The mysteries of Eleusis were not addressed to the reason, but to the eyes, to the imagination, to the heart." The greater part of the mystae carried away only confused impressions from these ceremonies, they did not find in them any moral instruction or philosophical revelation. The miracle-play, with its splendour and its pathos, was a performance to be followed with passionate interest, but whose meaning no one sought to discover; it touched the heart by its innate power. "I listened to these things with simplicity," says Plutarch, "as in the ceremonies of initiation and the mysteries, which do not allow of any demonstration or of any conviction produced by reason." Nevertheless, although there was no desire to help the worshipper to fathom the hidden meaning of the rites, or to explain to them their obscure symbols, the solemn ceremonies of Eleusis were not unfruitful. "The scenes which were unfolded before their eyes, the symbolical objects which they touched, the mysterious sentences they heard pronounced—all these rites, the exact sense of which they did not discover—were not without meaning for them, and awakened in their minds impressions of a higher kind than those which would have been aroused by a performance of a purely human interest." †

The noblest minds of antiquity, philosophers, states-

---

* Aristotle, *Fr. ed. Heitz*, p. 40.
† Decharme, *Myth.*, p. 399.

men, orators, historians and poets, professed the deepest reverence for the mysteries of Eleusis. From Pindar*
to Plato, from Isocrates to Cicero, they all join in acknowledging the profound influence they exercised over the minds of men. "It is said," writes Diodorus Siculus,† "that those who have taken part in the mysteries become more pious and upright, and better men in all respects, than they were before." Lucretius, too, who was not disposed to be indulgent towards the religions of ancient times, admits that the mysteries consoled the childhood of mankind: "*Primae dederunt solacia dulcia vitae.*" They cannot, then, have been mere pious observances and barren ceremonies, but they must have produced some wholesome effect upon the mind. The representation of the lower darkness where the goddess dwelt, the fairy picture of the joys of Olympus, awoke in the minds of the initiated some anxious thought for their own fate; in the midst of this dread gloom and of the heavenly light which bathed them in its brightness, they looked forward to the day when they themselves would descend into that lower world and would see those horrors and those joys. The thought of a future life, more than any other motive, drew the Greeks towards the sanctuary of Eleusis, and they found there both promises and hopes. The mystic poems recited in the sanctuary taught them that, after death, the initiated, crowned with flowers, would take their places at the feast of the righteous, there to taste of eternal joys, that they would form a company of the elect, seeing the gods

* Pindar, *Fr.* 102; Sophocles, *Fr.* 719; Isocrates, Pan. VI. p. 95, § 28; Cic., *Legg.* ii. 14; Plato, *Phaedrus*.
† *Hist.* v. 48.

face to face, and that in fields bathed in light, in an atmosphere which was always pure, in the midst of songs and games, they would enjoy eternal bliss. They learned, according to an inscription recently discovered at Eleusis, "the fair and joyful truth that death is not an evil but a blessing to mortals," and they carried away with them, according to Isocrates, "the brightest hopes, not only for the end of this life, but for all time to come." The profane, those who had not enjoyed the privilege of initiation, must look forward to eternal punishment after death, and Triptolemus himself sat in Hades to assign to the initiated the place of honour which was their due. "Thrice happy," cries Sophocles, "are those among mortals who shall go down into Hades, after having gazed upon those sacred rites, since for them and them alone life is possible in the lower world; for the rest there can be nothing but suffering." We see wherein the great merit of the mysteries consisted: by teaching the immortality of the soul, by making the happiness of this eternal life dependent on certain conditions of piety, purity and justice, by penetrating the soul with the feeling of the infinite, they contributed strongly towards maintaining the moral and religious traditions of humanity. Even in the last days of paganism, when the mysteries of Egyptian, Syrian or Phrygian origin were infected with the gravest disorders, the Eleusinian mysteries remained free from all corruption, and their teaching which was always philosophical rather than religious, preserved its character of austere gravity unimpaired until the day when, in 381 A.D., the sanctuary of Eleusis was closed for ever by the edict of Theodosius.

III.

The excavations have restored to us several of the more important buildings connected with the cultus of the great goddesses. In the northern corner of the sacred enclosure, whose walls and strong towers we can still trace, we find the larger Propylaea, with their six Doric columns and double front; the inner passage has Ionic columns at the sides, and a large medallion, which no doubt represented an emperor, crowned the outer pediment. Farther on are the lesser Propylaea, which formed the entrance to an earlier temenos, and within this, at the foot of the hill, on which the temple of Demeter was no doubt built, and where a chapel of the Virgin still conceals the remains of the ancient edifice, numerous buildings rose upon the plain. By the side of the Sacred Way, paved with marble and lined with statues, stood the temenos and the little temple of Pluto, built in the fourth century B.C., with the numerous grottoes which no doubt played some part in the cultus of this chthonic deity; close to the temple were the treasuries of the goddesses, one of which was perhaps situated on the small plateau reached by six steps cut in the rock; and farther on, against the southern wall of the peribolos, was the Bouleuterium, replaced in the Roman period by two sumptuous porticoes, between which was a secondary entrance into the temenos. The most important of the buildings, however, and that which first strikes the eye amidst the confused heaps of ruins, is the great hall of initiation, built in the fifth century under the direction of Ictinus. Its arrangement does not in any way recall

the ordinary plan of a Greek temple ; the portico of
twelve Doric columns, which now forms the façade, was
in fact added by the architect Philo in the fourth century,
and the original building had neither pronaos, cella, nor
opisthodomos. It is a large rectangular hall, more than
a hundred and sixty feet in length, and rather less in
breadth, entered by two doors on each of three sides,
and with eight tiers of stone seats running all round
the walls, on which nearly three thousand spectators
could easily find a place. The singular feature of this
hall, which was evidently intended to accommodate
a large number of spectators, was that the interior,
instead of being left free, was filled by a forest of
pillars arranged in six parallel rows which must have
entirely prevented the spectators, seated round the
walls, from seeing the performance as a whole. Incomplete explorations led archæologists at one time
to believe that there had been a subterranean shrine
here, or more probably rooms beneath the stage where
the machinery was prepared and the arrangements
made for the changes of scene and marvellous apparitions which in the miracle-play of Eleusis had such a
powerful effect on the imagination of the spectators.
Recent excavations have shown the incorrectness of
this hypothesis. The floor of the supposed crypt is on
the same level as the portico outside, and consequently
we are compelled to recognise in it the hall of initiation,
although we do not fully understand its object and
arrangement. A projecting piece of rock rises in the
centre of the hall; did it once support a colossal
statue, an altar, or some other object? We cannot tell.
All we know is that the building of the sixth century

which was replaced by the hall of Ictinus, though smaller in size, was on exactly the same plan. The excavations have been carried to a considerable depth, and have in several places laid bare portions of the building destroyed by the Persians, and shown that Ictinus only reproduced on an enlarged scale the hall of initiation designed by the unknown architect of the sixth century.

In front of this imposing building, supported by two powerful buttresses, there stretched between the portico of Philo and the temenos wall an extensive terrace, built over the remains of the old walls which had surrounded the sanctuary in early days. In this courtyard, paved with marble, stood numerous monuments, almost all dating from the Roman period, exedrae and bases of statues of which the dedications and ruins have been discovered. Among others there was a curious statue erected by a Roman, Q. Pompeius son of Aulus, and his two brothers, dedicated "to the power of Rome and to the eternal duration of the mysteries." Lastly, on both sides of the hall of initiation, broad stairs cut in the rock gave access to the upper terrace which lay in front of the temple of the goddess and above the buildings in the temenos.

The sanctuary of Eleusis, like the Acropolis of Athens, suffered severely, as we know, during the Persian invasion. Happily, Xerxes had not the tastes of a collector. He carried off little or nothing from the works of art of the conquered people, and when the storm was over the Greeks found in the midst of the ruins the mutilated remains of the statues dedicated to their gods. We have seen, from the account of the excavations on the Acropolis, that they did not think it worth while to replace these broken and headless statues on their bases, but that, covering these relics of the past with a thick layer of earth, they built new and more beautiful temples above the former shrines. They did at Eleusis as they were doing at Athens, and we cannot rejoice too heartily over this good fortune. Whilst the celebrated statues of the fifth and fourth centuries B.C. have disappeared for the most part in the shipwreck of paganism, the earth has faithfully sheltered these treasures which were disdained by the Greece of Pericles, and which possess such incalculable value for us. Below the level of the temples of Eleusis the excavations have brought to light precious relics of archaic art, and although these marbles are few in number, and cannot be compared to the Acropolis statues, and although none of the bronzes have been discovered here which were so plentiful at Olympia, yet the exploration of the shrine of Demeter has enriched the history of ancient art with some noteworthy examples.

It is not necessary to discuss at length the female statues reproducing the type with which we are already acquainted from the excavations at Delos and on the Acropolis. Here too we can follow the slow and steady progress of early Greek art, from the almost shapeless figure, still encased in its stiff sheath, up to the nearly perfect statues which are the works of ripened archaism. All the stages in this long journey are marked by some statue at Eleusis : first comes the old *xoanon*, scarcely blocked out, encased in its sheath from waist to feet, the bust alone expanding rather more freely under the heavy folds of its drapery; then the female statue with the long tunic falling in folds and

finely folded upper garment, the wide mantle over the
shoulders, the carefully dressed hair, and calm, hieratic
attitude. Two among the figures of this type found at
Eleusis are especially interesting, for they belong to
that period of transition in which archaism, as it draws
to a close, is freeing itself from conventionality, and
making an effort after fidelity to nature. One, in
accordance with the conventional method adopted by
early artists to indicate motion, has one leg advanced
and the other bent; while the body faces the spectator,
the legs are seen in profile; the draperies, however,
are already treated with marvellous delicacy, and a
pliant and youthful form is visible under the different
stuffs which the artist has distinguished with rare
skill. It is still faithful, it is true, to archaic tradition,
but the technique is already remarkable. The other
statue is more original, and shows more of the sculptor's
individuality; the hair is treated with more freedom, the
attitude is more supple, the novel arrangement of the
drapery betrays an ardent desire for something newer
and better, while the nude portions show a more careful
study of anatomy. It is true that the extremities are
wanting in the delicacy and refinement of Attic work,
but we are sensible of a firm resolve to break at last
through the monotonous conventions of a school, and
this renders the statue worthy of attention. The head
is unfortunately missing in all these figures; but some
archaic heads have been discovered at Eleusis which are
not unlike the less beautiful among the Athenian heads:
they have the same symmetrical arrangement of the hair,
which is dyed red and surmounted by a blue diadem, the
same eyes inclined upwards towards the temples, the

same mouth with its half smile, and the same sketchy modelling which does not render with exactitude any of the details of the too precise oval of the face.

As to the artistic riches which adorned the temple of Eleusis after it had been rebuilt, they have almost all disappeared. Scarcely anything remains but a fragment of a colossal statue of the Roman period, which is now in the Fitzwilliam Museum at Cambridge, a number of rather poor bas-reliefs found in the temenos of Pluto, and two pieces of sculpture both of which are worthy of attention. One is the celebrated bas-relief in the museum at Athens, in which Demeter in the presence of her daughter is giving the young Triptolemus the ear of wheat which is to make the earth fruitful; it is one of the most remarkable productions of Attic art in the period immediately preceding Phidias, and the delicacy of the modelling, the pliancy and grace of the movements, belong to the purest style. The other is a beautiful head of a young man, discovered in 1885 near the sanctuary of Pluto, which is certainly one of the finest pieces of Greek sculpture of the fourth century extant.

An inscription found near the statue bore a dedication to Eubuleus, a divinity whose cultus was in the fifth century associated in the religious worship of Eleusis with that of Triptolemus and the divine pair mysteriously called "the god and the goddess." According to the legend, this personage was a brother of Triptolemus, who was feeding his flock upon the plain at the time of Cora's abduction, but, as a matter of fact, the name designated, by a kind of euphemism, a chthonic deity; as Pluto means the god who enriches, so Eubuleus signifies the beneficent. This was one of the forms of the chthonic divinity worshipped at Eleusis, who never assumed in this cultus the savage and destructive character of Hades. Closely associated with the great goddesses, Pluto-Eubuleus watched with kindly care over the seed buried in the soil, and brought the harvest forth out of the earth which is his kingdom. The worship of this deity, which had been cast into the shade by that of Iacchus, seems to have undergone a kind of revival in the fourth century, and the temple where the head was found dates precisely from that period. We may then conclude without improbability that this head, with its dreamy and melancholy expression, represents Eubuleus himself, and this hypothesis, by becoming a certainty, has at the same time shown that this marble was an original work by one of the greatest sculptors of antiquity.

It is well known that Praxiteles chose the motive of several of his statues from the cycle of Eleusis, and made for the sanctuary figures of Demeter, Cora, and Dionysus, as well as a group representing Triptolemus and the great goddesses, and another well known in ancient times under the name of the Katagousa, which represented Demeter carrying her child. The somewhat melancholy mythology of Eleusis seems to have had a particular fascination for this master, and to have gratified his desire of modelling expressive statues. He also sculptured an Eubuleus, and this work was no doubt placed in the sanctuary of Eleusis. The style of the recently discovered head is undoubtedly that of a great master; it is, on the other hand, unquestionably an original which was famous in ancient times, for copies of it exist at Rome and Mantua; lastly, the extremely skilful technique of the work, the delicacy and care with which the hair is treated, and the shape of the forehead, strongly recall the Hermes of Olympia. We may then safely recognise in this marble a work of Praxiteles.* The head is undoubtedly very different from that of the Hermes, the expression has more individuality, and its gravity is tempered by infinite sweetness, but the Hermes is a work of the master's youth, while the Eubuleus marks the full maturity of his genius. Unfortunately the marble has suffered, and the breakage has impaired the effect of the work. It is none the less a valuable discovery, and it is not the least merit of the excavations of Eleusis that they have restored to us an original work by the great Athenian master of the fourth century.

* Dr. Otto Kern (*Mitth.*, vol. xvi.) denies that the head is the work of Praxiteles, and declares that it represents Triptolemus (the name Eubuleus having no connection with it, and being in fact a title of Zeus). —E. R. P.

CHAPTER IX.

THE EXCAVATIONS OF EPIDAURUS (1881-87).

BOOKS OF REFERENCE:—
    Professor Gardner, "New Chapters in Greek History."
    Louis Dyer, "The Gods in Greece."
    Cavvadias, "Rapports sur les fouilles d'Epidaure" (*Praktika* 1881-85).
    Staïs, *Ibid.* (1886-7).
    Cavvadias, "Inscriptions d'Epidaure" (*Ephemeris Archaiologike*, 1883-6).
    Staïs, *Ibid.* (1887).
    Cavvadias, "Monuments d'Epidaure" („   „   1884-5).
    Staïs   „   „   („   „   1886).
    S. Reinach, "Chronique d'Orient" (*Rev. Archéol.*, 1884).
     „   „   "La Seconde Stèle des Guérisons Miraculeuses" (*Rev. Arch.*, 1885).
     „   „   "Les Chiens dans le Culte d'Esculape" (*Rev. Arch.*, 1884).
    Petersen, "Athenastatuen von Epidauros" (*Mitth.*, vol. xi.).
    Furtwaengler, "Epidauros" (*Berliner Philol. Wochenschrift*, 1888, p. 1484).
    Defrasse et Lechat, "Notes sur Epidaure" (*Bull. Corr. Hell.*, 1890).
    Foucart, "Sur les Sculptures et la date de quelques Édifices d'Epidaure."
    Girard, "L'Asclépieion d'Athènes." Paris, 1881.

AMONG the religious cults of ancient times that of Asclepius, whom the Romans called Aesculapius, was certainly one of the most interesting and remarkable, as it is in some respects also one of the most modern. Whatever liberties Hellenic piety may have taken at times with the other Olympic deities, with Zeus for

example or Apollo, the prayers addressed to them by
their worshippers were mainly spiritual and moral in
their nature; these deities were personages of too much
importance to be readily appealed to in all the petty
miseries of mankind. The case was quite different with
Asclepius. He was a physician, and we all know that we
keep no secrets from our doctor; thus his worshippers
revealed to him all the troubles and annoyances, not of
their souls, but of their bodies; at the foot of his altars
they did not scruple to display all the infirmities of
human nature; their piety was quite at its ease with
him, and in dishabille, so to speak. The study of the
cultus of this god of help and healing, whose temples
seem to us to have been really hospitals, the only
charitable institutions ever originated by the ancient
Greeks, forms one of the most interesting chapters in
their religious history. Thanks to the zeal and devotion
of the Archæological Society of Athens, we can easily
pursue this study to-day; the excavations conducted by
them in 1876-77 on the southern slope of the Acropolis
made us acquainted with the precinct of Asclepius at
Athens, and shortly after, the important explorations at
Epidaurus, carried on from 1881-87 under the direction
of M. Cavvadias, brought to light the most famous and
splendid of his temples, and restored to us a great
number of monuments of all kinds which throw quite
a new light on a cultus hitherto almost unknown.

I.

A traveller of the second century A.D., who saw the
dying splendour of Pagan Greece, has given us a somewhat lengthened description of the wonders of the temple

of Epidaurus, and has explained precisely what were
the essential elements of every sanctuary of Asclepius.
"Three things were included in it: a temple which
sheltered the statue of the god, porticoes or broad and
airy galleries where his guests found temporary refuge,
and lastly a spring which provided the water necessary
for the treatment which he prescribed, and for the
purifications and ablutions of the suppliants."* In the
greater number of the temples the buildings were
very simple: the important matter being the convenient
arrangement of the religious hospital attached to the
temple, it was necessary, as M. Girard remarks, to secure
large empty spaces for the construction of the porticoes
destined to receive the patients, together with spacious
courtyards and easy means of ingress and egress, in
order that the crowd of pilgrims might move about
with ease; so that in order to accommodate the
dependants of the god, his temple was encroached
upon. Nor was it, as a rule, one of the splendid
buildings erected at great cost, with the aid of the most
illustrious artists of Greece; Asclepius contented himself with a modest chapel, where valuable offerings and
curious ex-votos offered by the grateful piety of his
worshippers were collected around his statue. The
greater part of the sacred precinct was occupied by
wide, airy colonnades, where the crowds of worshippers
were lodged; near these was the miraculous spring,
and sometimes a sacred wood as well, which diffused a
coolness and shade beneficial to the sufferers. Such was
also the case at Epidaurus; but on the other hand, here,
at the most celebrated of all the temples of the god, the

* Girard, *L'Asclépieion d'Athènes*, pp. 5, 6.

aspect of the buildings was far more splendid. Within the sacred precinct, entered through propylaea of the Doric order, there rose some buildings of considerable size, of whose arrangements we now know the most important details. There was the temple of Asclepius, nearly 80 feet long by about 40 feet in width, with Doric columns and painted cornice; its pediments were decorated with statues, and figures were placed upon the summit and at the angles of the principal façade; it was built in the early part of the fourth century, and strangely enough had no opisthodomos or chamber to the rear of the cella. Near it stood a fine circular building, the Tholus, built in the fourth century by Polycleitus the younger, a splendid marble rotunda, surrounded by a Doric colonnade, and ornamented on its inner circumference by sixteen Corinthian columns. This was one of the most beautiful buildings at Epidaurus, with its coffered ceilings painted in a thousand different colours, with its walls decorated with all kinds of marbles, and lions' heads and marble ornaments crowning its summit; it was also one of the most remarkable, with its circular crypt, where three concentric passages no doubt formed a reservoir fed by the sacred spring. Both buildings contained masterpieces of art: in the temple was the gold and ivory statue by Thrasymedes of Paros, representing Asclepius seated on his throne, and accompanied by his sacred animals, the serpent and the dog, frequently connected with the cultus and the miracles of the god; while in the tholus, entirely decorated with paintings, was to be seen among others the famous Methe of Pausias, a woman drinking out of a cup of glass, much praised by the

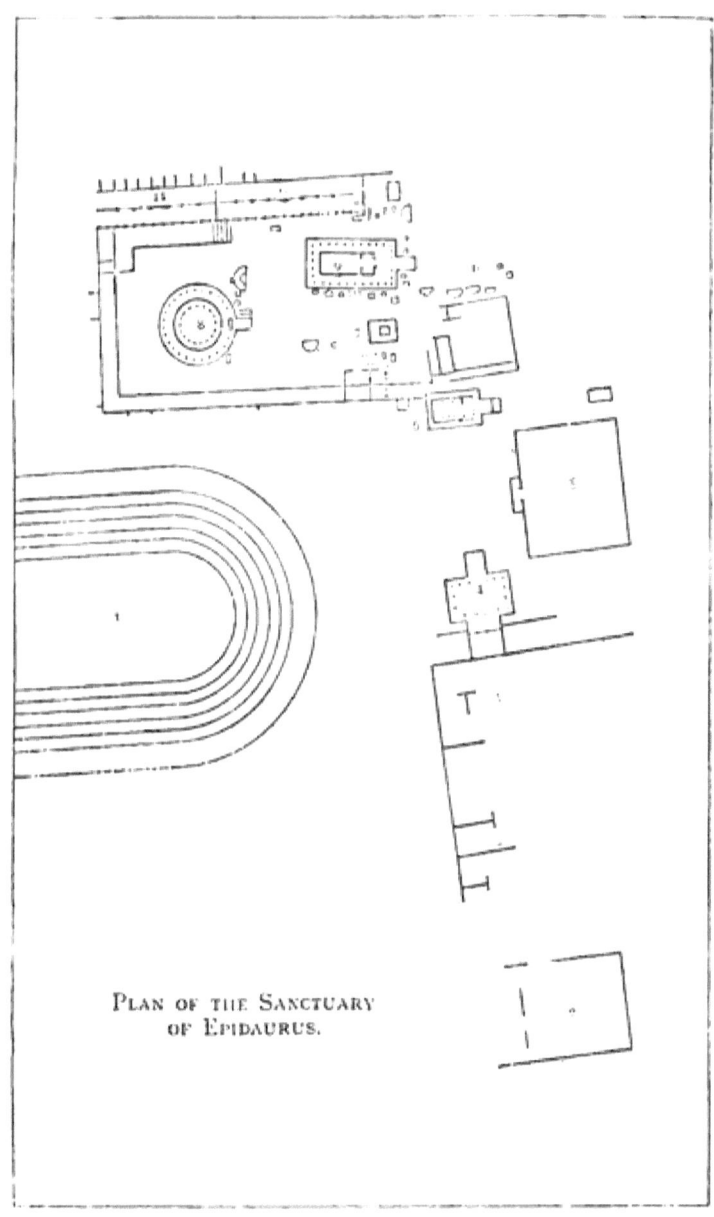

PLAN OF THE SANCTUARY
OF EPIDAURUS.

1. Stadium.
2. Temple.
3. Gymnasium.
4. Propylaea.
5. Large building, use unknown.
6. Temple of Artemis.
7. Altar.
8. Tholus of Polycleitus.
9. Temple of Asclepius.
10. Statues.
11. Double portico.
12. Portico.

ancients for its wonderful transparency. A little farther on was the Doric temple of Artemis, in front of which there was an altar; the goddess was worshipped there under the name of Hecate, and as a divinity of healing and succour her cultus was closely connected with that of Asclepius. Then came the chapels built in honour of Hygieia, the beloved daughter of the god, of Aphrodite and of Themis, and of Apollo Maleatas, who at Tricca, his Thessalian shrine, worked miraculous cures like Asclepius, and to whom the faithful often addressed their prayers and offered their worship before going to Epidaurus. In the midst of the precinct stood a large altar, and along its northern wall there ran two Ionic porticoes, one of which consisted of two stories, one above the other, which formed the sacred dormitory or abaton, divided into two long bays, where the patients spent the night; at the eastern end of this building was a well, from which came the water needed for the treatment prescribed and for purifications. There were many other buildings as well: a large square edifice to the south of the temple of Asclepius, which was found full of statues and offerings; another, of which the purpose is unknown, between the temple of Artemis and the Propylaea; to the north of the Abaton magnificent baths, adorned with statues and built by the emperor Antoninus, in which a library has been found among the many halls grouped round a wide courtyard; lastly, outside the precinct lay the stadium, and at a greater distance on the opposite hill-side the theatre, also built by Polycleitus the younger, which was considered one of the largest and finest in Greece. The excavations have restored to us this magnificent edifice in an almost perfect state of preservation, with its thirty-two tiers of seats, its circular orchestra, of which the centre was occupied by an altar, and its wall decorated with columns and statues supporting the stage; while before it the view stretched far away over the beautiful picture formed by the buildings within the sacred precinct. Everywhere, on the terraces of the peribolos and around the temples, there rose a multitude of statues representing the god, richly-decorated exedrae, monuments erected to famous doctors, stelae or tablets recording the miraculous cures worked at the sanctuary, and pious ex-votos and bas reliefs painted in bright colours, bringing before the visitors the miracles of the god or the homage paid by the worshippers. Only one thing was wanting to this sanctuary of the healing art: one might be cured there of any disease, but one could not die in peace. It was forbidden to desecrate any holy place, Epidaurus as well as Delos, with the pollution inseparable from birth and death; and the sick, when on the point of death, were mercilessly driven out of the sacred precinct, whither sometimes they had come from the most distant parts of Greece to consult the god, and sank down overcome by exhaustion and fatigue on the very threshold of the temple. It was left for the benevolence of a Roman Emperor to put an end to this wretched and barbarous state of things. Antoninus had a large building erected outside the boundaries of the peribolos, in which the dying might at least pass their last moments in peace. Such refinements of humanity were always deemed superfluous in Greece.

Such is the scenery which formed the background

of the festivals of Asclepius, and amidst which the superstitious crowd of worshippers came every day to the foot of the altars to lay bare their sorrows and to await the marvellous effects of the almighty power of the saviour-god. It is time now to bring forward the principal characters in the play.

Asclepius, like every other deity, had his high priest, whose duties did not differ much from those of the ministers of other cults: the charge of the sacrifices, the superintendence of the temple decorations, the management of the sacred treasure, formed the ordinary business of every priestly body. Still it seems likely that the character of the god in whose service he stood, imposed certain special duties upon the high priest of Epidaurus. The crowds of sick persons who came to the temple in search of health must undoubtedly have imposed various obligations upon him. It must have been necessary for him to exercise some supervision over the multitudes crowded together in this vast caravansary, and undoubtedly the duty of keeping order within the sanctuary devolved upon him: in any case the head of the priesthood at Epidaurus could not be entirely indifferent to the pilgrims who flocked to the temple, for the more frequent and wonderful the cures, and the more numerous the patients, the more glory and profit accrued to himself; consequently the high-priest exercised absolute control over the sacred hospital entrusted to his care, although he does not seem to have operated himself. It was the inferior officials of the temple who came into direct contact with the sick, whose duty it was to receive them, to instal them in their places in the porticoes, to bestow certain elementary attentions upon them, and to see that the prescriptions of the god were properly carried out. Among these subordinate functionaries the inscriptions name the zacorus, the cleiduchus or key-beaver, the pyrphorus or torch-bearer. Their duty was simply that of acting as intermediaries between the god and his worshippers; it was not from them that the cure came, and Asclepius himself received all the honour. The prescriptions given by the god were communicated to the priests, and it was their business simply to see that the miracle which had been foretold did not fail; consequently it is rare to meet with a doctor among them—whose presence indeed must have been quite unnecessary, when we think what the course of treatment at these temples really amounted to. Certainly medical science reached a very high point in ancient Greece: the schools of Cos and Cnidus, of Rhodes and Cyrene, were famous throughout the Greek world, but the doctors trained in them were mostly laymen. In fact by the side of the religious treatment practised in the sanctuary, every city had lay doctors, and by the side of the religious hospital, which lay under a slight suspicion of charlatanism, the city organised a civil hospital under the direction of a physician appointed and paid by the state. In this building was to be found all the equipment necessary for scientific treatment—a consulting room, an operating theatre with all its instruments and apparatus, levers of all kinds, a beam from which to suspend the patient by the feet in order to reduce dislocations of the thigh, knives and bistouries, trephines and instruments for cutting the uvula, premises for the preparation of drugs, and lastly, rooms for the sick. In establishments of this kind the treatment was thoroughly scientific, carried out under the eye and by the hands of a professional physician. In the temples of Asclepius nothing of the kind was to be found. The variety of ailments and the recurrence of cases of the same kind no doubt gave the priests at last a certain amount of practical experience, but their share in the treatment was limited as a rule to certain simple attentions to the sick. The cures effected were the purely divine and supernatural effects of the intervention of a higher power. "A sick man who came to take up his place under the porticoes of the temple of Asclepius, and who went to sleep there, after having offered the customary sacrifices and gone through the usual formalities, in the hope that the god would appear to him and point out the remedy for his disease, did not rely for his cure upon the aid of human science; he put his whole trust in the working of a higher power,—it was a miracle that he expected. We can therefore understand how it was that the presence of a doctor was unnecessary; all that was wanted was for the priest and his subordinates to see that the miracle did not fail. The Asclepieum was not a hospital to which one went to be treated by men who had enjoyed professional training and long practice; it was a temple to which men resorted in order to place themselves under the protection of a divinity."

II.

Pausanias relates that at Epidaurus he saw six tablets recording the miraculous cures effected by

Asclepius in his sanctuary, and giving, together with
the name of the sick person, a statement of the disease
and of the treatment prescribed by the god. By a rare
piece of good fortune two of these interesting records
have been found in the course of the excavations. Two
very long inscriptions, both dating from the fourth
century, have been discovered near the large Ionic
portico, and make us acquainted in a very curious
fashion with the sacred therapeutics. Each inscription
mentions about twenty cases which give an admirable
idea of the tact and skill of the divine physician. We
will only briefly mention the cases of the one-eyed or
the blind, whose sight Asclepius restored by rubbing
the empty socket with a salve of his own making, or of
the lame whose crutches were stolen during their sleep,
and who in the morning could run after the thief; we
will say nothing, interesting as the case is, of that man
of Mytilene who, as the inscription says, "had no hair
on his head but plenty on his chin," and for whom
Asclepius made a thick head of hair grow in a single
night; we will not even dwell upon the paralytics
whom the god commanded to go out and to bring a
heavy stone into the temple,—these are but the "small
change" of the miraculous cures. We will hasten on to
the more remarkable cases—the treatment of dropsy,
for example. There was a Spartan woman who was
suffering from dropsy. Her mother went to consult the
god of Epidaurus on her account, for Asclepius could
work his cures from a distance, and the presence of the
sick person was not indispensable. She fell asleep
within the precinct, and had the following vision: "It
seemed to her that the god cut off her daughter's head

and hung up her body neck downwards; the water ran out in streams, and then he took the body down and replaced the head upon the neck. After she had seen the vision the mother returned to Lacedaemon and found her daughter restored to health." Let us see how Asclepius treated a case of cancer. There was a man who had a cancer in his stomach; he went to Epidaurus, slept, and had a vision. "It seemed to him that the god commanded his attendants to seize him and hold him firmly while he opened his stomach. The man began to run away, but the attendants caught him and bound him; then Asclepius opened his stomach, cut out the cancer, sewed up the incision, and released the man from his bonds." Immediately he went away healed. The god had equally efficacious remedies against gout and headache; he restored speech to the dumb, health to those who were suffering from incurable wounds, ease to those afflicted by leprosy or vermin; he heard the prayers of barren women, and of those who sought for delivery when too long-delayed; he knew the remedy for every ailment, and had prescriptions for all kinds of sickness, and it is to be regretted that the coarseness of the details prevents our giving the particulars of some of his cases. He was certainly in favour of violent treatment: to cure a bandy-legged man he laid him on the ground in front of the temple, and mounting his chariot, he trampled him under the horses' hoofs until his legs were straight. He was an ardent lover of surgical operations. One of his patients had swallowed leeches: what did the god do? "It seemed to this man," says the inscription, "that the god cut open his chest with a knife, and taking out the leeches, gave them into his hands and sewed up his chest again. When day broke he departed, carrying the leeches in his hands, and from that moment he was cured." The cause of this malady is scarcely less remarkable than the treatment. It was due, according to the inscription, to a trick of the patient's stepmother, who had put the leeches into a mixture of wine and honey which he had swallowed. We see how cruel were the sufferings to which the stepsons of ancient times were exposed.

One of the most wonderful cases, and one of the most celebrated in antiquity, was the miraculous cure of a woman who suffered from a worm of monstrous size. Her case was given up by the doctors. She went to sleep in the temple of Asclepius at Troezen, and it seemed to her that the sons of the god—who was too much engaged to come himself—cut off her head, plunged their hands into her intestines, and pulled out the worm; but, when they had done so, they could not put the head on the poor woman's body again. They had to send in hot haste to Epidaurus to beg Asclepius to come and repair the effects of their clumsiness. The god hastened to their aid, reproved the imprudence of his children, who wished to do more than their skill admitted of, and then with his unrivalled art, put the head on the shoulders again, and the woman was healed.

Asclepius did not give his services for nothing; he made a point of receiving his fee, and sometimes it was a high one, for we are told of one remarkable cure for which he charged nearly £2500. He was not to be trifled with in this matter, and did not hesitate to use the irresistible means of compulsion which his divine character put into his hands. One blind man whom he had healed refused to pay the price of his cure, and was at once deprived of his sight again; but when he came humbly promising to be more grateful in future, Asclepius was generous enough to heal him once more. He was always a good-natured god, who was not above being merry, and liked a joke. A sick child promised him ten knucklebones as a fee, and the offer amused Asclepius so much that he cured his little patient at once. He made merry in a very irreverent fashion at the cost of a good woman who had put the questions she asked of the god in a clumsy way; he bore no grudge against the sceptics who mistrusted his power, but only tried to convince them of the injustice of their doubts. A man whose fingers were paralysed came one day to Epidaurus, and seeing the votive tablets set up to commemorate miraculous cures, began to ridicule all these inscriptions. What did the god do? He appeared to the sceptic in a dream, and stretching out his fingers one after the other, restored him the use of his hands; then, when the man in his surprise could not credit the miracle, and was clenching and extending his fingers in turn, Asclepius came back to him and asked him if he were still sceptical. He replied that he was not. Then the god said, "Since but just now thou didst not believe things which are not incredible, I will now bestow upon thee an incredible cure." This was the only vengeance he took. The same thing happened to a woman from Athens. "Walking round the precinct, she began to ridicule some of the cures, saying that it was absurd and impossible for the lame to walk and the blind to see, simply because they had had a vision. And sleeping, she saw a vision: she thought the god stood by her and said that he would make her whole, but that he required her as payment to dedicate in the precinct a silver pig, as a memorial of her folly. And when the day broke she went away healed." In short, the god was thoroughly good-natured, kindly, and obliging, always ready to work a miracle, even when it was quite foreign to his craft, and making success in difficult cases a point of honour. Here is the case of the porter who let his bag fall, and broke the cup out of which his master used to drink. As the man, much grieved, was trying in vain to fit the pieces together, a passer-by saw him, and said with a smile, " Why, unfortunate, do you try in vain to put the cup together? Asclepius himself, the god of Epidaurus, could not make it whole." This was enough: the god thus challenged was on his mettle, and when the slave, on reaching the temple, opened his bag, he found the cup miraculously mended. Nothing was too hard for Asclepius; he could mend broken crockery as easily as broken limbs, and was as ready to work a miracle to save Sparta from the attacks of Philip of Macedon as to restore to health the sick who sought his aid.

From these curious tablets of Epidaurus we can get an admirable idea of the ordinary methods of treatment in use at the shrine. There is no question of pharmaceutical remedies; the inscriptions only speak of visions and dreams.* It was not by ordinary remedies, but by

---

* There is, however, on the second tablet, a curious case which seems to show that the priests sometimes performed actual surgical operations after having sent the patient to sleep: cf. Reinach, *Chroniques d'Orient*, Paris, 1891, p. 93.

purely supernatural intervention, that the god relieved the sufferings of the faithful who sought his aid. The maladies, too, to which he devoted himself belonged to the province of miracle rather than to that of science, and professional physicians would as a rule have found as much difficulty in treating them as in curing them. This is not the least curious side of these stories which throw such interesting light on the therapeutics of ancient temples.

We can easily imagine that these miraculous cures, of which the fame spread throughout the Greek world, attracted many patients to the temple of Asclepius; and as one could go there not only on one's own behalf, but on that of others, if the sick people were too seriously ill to undertake the pious pilgrimage in person, the sanctuary was always full. Every class in society furnished its contingent; rich and poor, great and small, all met there at the foot of the altars, but the greater number were women. The worship of Asclepius was in fact well suited to attract them: the dreams, the nocturnal revelations, the miracles, all the attendant circumstances, by inspiring them with a holy terror, charmed their imagination and attracted them to the god. Religious bodies placed themselves under the protection of Asclepius, fanatical devotees passed their whole lives under the shadow of the sanctuary, inflamed by a kind of mystical love for the god who had bestowed some favour upon them, and even the laymen of the medical profession, although they were in direct competition with the unlearned practitioners of the Asclepieia who robbed them of many patients, lowered their colours, if not before the priestly wonder-workers whom they suspected of quackery, at least before the god whose almighty powers they recognised. Public physicians willingly placed themselves under the protection of the god of healing, and thanked him for the fortunate cures they had accomplished. The inscriptions prove this subordination of science to religion, and the writings of Hippocrates bear many traces of it: he knows no other remedy for epilepsy than sacrifices, vows and supplications, and as a cure for melancholy he recommends the worship of the gods. Elsewhere, too, there occurs this remarkable passage: " As regards diseases and their symptoms, medicine in the greater number of cases inclines to do honour to the gods. Physicians bow before them, for medicine has no superabundance of power."* Thus faith in the supernatural existed even in men of science; medicine often confessed itself powerless before religion, and the appeal to divine assistance always remained, says M. Girard, for the learned as well as for the ignorant, the final resource and the last hope.

A curious scene in the *Plutus*† of Aristophanes, in which, through the exaggerations of comedy, it is easy to distinguish the elements of truth, describes to us with great exactitude the way in which one of the pilgrims, who came in crowds to Epidaurus to pray for healing, spent his day. When the sick man was brought to the temple, it was customary first to purify him, by plunging him into the sacred spring. Worshippers were to present themselves pure before the God; and purity meant for the ancients, as it does

* Hippocrates, Περι ἐυσχημοσυνης.
† *Plutus*, 653, 899.

for the Orientals of our own time, corporal cleanliness* more than anything else; in a hospital, moreover, such as the sanctuary really was, precautions of this kind were necessary measures of hygiene, and the priests no doubt were careful to see that no one took up his place in the porticoes without first having carefully bathed. Then, having consecrated some cakes and other delicacies on the altar of Asclepius, or offered to the god, if one were richer, the more costly sacrifice of a pig or a ram, he took up his position for the night in the sacred dormitory. The patients had to bring with them all they needed, both their provisions and the coverlets in which they wrapped themselves, for the authorities of the temple only put at their disposal modest beds of straw. At nightfall they lighted the sacred lamps, and the priest recited a kind of evening prayer, to entreat the divine protection for the pilgrims. Then they all lay down, and the temple-servant went through the porticoes putting out the lights, and bidding the sick people go to sleep. The holy night had begun.

When the worshippers had fallen asleep, their imaginations inflamed by the expectation of some divine vision, and their minds excited by the atmosphere of the temple and the solemn surroundings of the evening prayer, Asclepius appeared to them in a dream, and revealed to them either the treatment they were to adopt or the religious observances they were to perform in exchange for a sudden and miraculous cure. We have had some examples of the miracles which the god worked as he went his way; the prescriptions he gave

* Girard, *L'Asclépicion*, p. 71.

were not less remarkable. To some he recommended baths ; some he advised to drink lime-water or hemlock-juice, and for others he prescribed gymnastic exercises and cold baths. One of his directions was to eat a spiced partridge ; another, a sovereign remedy against pleurisy, was to apply to the side affected a poultice of ashes moistened with wine. A certain cure for spitting blood was to eat fir-cones cooked with honey ; to recover one's sight, nothing was better than a certain eye-salve made with the blood of a white cock ; lastly, water, played a more important part in his treatment than anything else, both applied externally by means of baths, or internally as a drink."*

All these prescriptions were very simple, as we see ; if the treatment he administered with his own hands was somewhat rough, Asclepius prescribed little to his worshippers beyond simple hygienic measures ; the rest was a matter of faith, and the miraculous intervention of the god always played the principal part in the cure.

In the morning, all the patients related to the priests the directions which Asclepius had communicated to them in their dreams, and busied themselves in carrying out the divine prescriptions. They hastened to the spring, and waited eagerly for the promised miracle. When sudden cures took place, there were cries of joy, and endless congratulations addressed to the fortunate recipients of the blessing ; but often the revelation was awaited night after night, the god put the patience of his worshippers to the proof, and was slow to appear ; still each new cure brought fresh hope with it, and gave to those whose prayers were still unanswered,

* Girard, *L'Asclépieion*, p. 71.

strength to wait until the day should come when the god would take pity on them too.

The influx of visitors was still greater at the time of the great festivals celebrated in honour of Asclepius, the Epidauria and Asclepieia. Then the temple was magnificently decorated, the statues were adorned with their finest jewels, the tables loaded with the sacred banquet offered to the god ; then, too, took place those holy vigils when all night long, by torchlight, the worshippers, their souls filled with the hope of a blessing, addressed their prayers and invocations to the god ; and then, among this great multitude, in this overheated atmosphere, were no doubt worked the most wonderful of the miracles, those which most powerfully impressed the imagination of the people. At such a time the precinct was full of noise and stir ; together with the multitude of worshippers doing homage to the god, and of sick people, accompanied often by their whole families, who had come to pray for their restoration to health, a crowd of spectators and an army of merchants thronged around the temenos. Dealers in ex-votos and curiosities set up their stalls in the plain, and to this great fair held under the shadow of the temple, worldly needs and pleasure attracted as many visitors as piety and devotion led worshippers to the foot of the altars.

When once the blessing had been received, it remained for the worshipper to discharge his debt of gratitude to the god. On this point ancient piety was of a very practical nature, and made a real bargain with the divinity, a reciprocal arrangement which gave nothing for nothing. Propitiatory sacrifices intended to conciliate the good-will of the god were very rare, and always very modest; a man did not pay his debt until his prayer had been heard. If Asclepius remained deaf to the entreaties of his worshipper, he lost thereby all right to the promised offering; on the other hand, if he was faithful to the terms of the agreement, the devotee was exact in payment, and as Asclepius was a deity who kept his word, popular piety was lavish of gifts and ex-votos. Sometimes the god, in the vision in which he appeared to the suppliant, himself named the offering he would prefer, and the inscriptions show that in this case Asclepius, with remarkable courtesy, usually directed the faithful to the altars of other divinities. Apollo Maleatas and Athena Polias, Hygieia and the sons of Asclepius, Helios and the Dioscuri, Telesphorus the saviour and the mother of the gods—sometimes indeed the whole Greek pantheon—received these attentions paid by the god of Epidaurus at the expense of his worshippers. At other times he allowed his worshippers to discharge their debt in what way they pleased, and then the variety of offerings was infinite. Sometimes they took the shape of painted reliefs representing scenes of worship or prayer, showing the god at the bedside of the patient, or depicting the precise moment of the miraculous cure, somewhat like those naïve pictures hung round the necks of the blind, which bring before our eyes the catastrophe in which they lost their sight. Sometimes they were statues or portraits of the sick who had been healed; most frequently they consisted of a multitude of small articles in marble or metal, the list of which, preserved in the temple inventories, is very curious. Some offered to the god a reproduction on a small scale of the body of the person healed, others consecrated a representation of the part affected, in

Ex-voto to Asclepius (Athens).

memory of the benefit they had received. There were whole faces, mouths and eyes, noses and ears, hands and feet, legs and trunks, hearts and breasts in gold or silver, or simply in gilded metal, to suit more modest purses,—just as in Italy to-day we find offerings of this kind suspended on the walls of famous shrines. Then there were surgical instruments, probes, medicine-chests, the gifts of physicians to Asclepius; there were vases, toilet-requisites, clothes, shoes, jewels, all the thousand things that the gratitude of the faithful consecrated in the sanctuaries of antiquity. Lastly, there were metrical inscriptions, hymns, pæans, sometimes composed by the sufferers themselves, and engraved in the sanctuary in honour of the god. One of these poems has been found at Epidaurus, in which a certain Isyllus,[*] at the command of the oracle of Delphi, sang the genealogy and the miracles of Asclepius, and elsewhere there are thanksgivings which reveal a naïve and touching piety. "Listen," says one of these poets "to the words of thy faithful zacorus: O Asclepius, how shall I go into thy gilded dwelling, O god, blessed and desired, beloved, how shall I go, deprived of those feet which once led me to thy sanctuary, if thyself, in thy goodness, thou dost not take me there after having healed me, that I may behold thee, O my god, whose glory surpasses that of the earth in spring? Here is the prayer of Diophantus. Save me, and heal my grievous gout, O blessed and most mighty presence. I adjure thee by thy father, to whom I loudly pray. None among mortals can bring a surcease from such pangs; thou alone, divinely blessed one, thou hast the power, for the supreme gods bestowed thee upon men, inestimable gift! to take pity on their sufferings and relieve them. O blessed Asclepius, god of healing, it is thanks to thy skill that Diophantus, relieved of his

---

[*] Ephemeris, 1885, p. 74.

incurable and horrible disease, will no longer go like a crab ; he will no longer walk upon thorns, but he will have a sound foot as thou hast decreed." *

III.

It was towards the end of the fifth century that the sanctuary of Epidaurus seems to have reached the height of its fame. It was then, almost at the same time that the temples of Sunium, of Rhamnus and of Phigalia were being built in other parts of Greece, that the sacred precinct of Asclepius was filled with splendid edifices ; the Tholus and the theatre, the colonnades of the Abaton and the temple of the god, date from this period, and the excavations have shown with what magnificence all these buildings were carried out. For the temple especially no expense was spared ; a long and interesting inscription found at Epidaurus has made us acquainted with all the details of its construction, and has given us, in the amount expended on the building, a presentiment of its splendours. The works of art discovered in the course of the excavations confirm this testimony, and give a very high idea of the sculptural decoration of the sanctuary.

A large number of statues in Pentelic marble were found in front of both façades of the temple ; they undoubtedly belonged to the pedimental sculptures, and although most of them are very much injured, it is by no means impossible to reconstruct the composition of which they formed a part. On the eastern pediment, the scene represented was the battle between Centaurs

---

* The translation is taken from M. Girard, *L'Asclépieion*, p. 121-122. See Prof. Dyer's *Gods in Greece*.—E.R.P.

and Lapiths, of which we possess only a few fragments —amongst others a fine head of a Centaur; on the western pediment appears the same subject which we find at the temple of Phigalia and on the mausoleum of Halicarnassus, the struggle between Greeks and Amazons. Many of the pediment figures have come down to us almost intact, and we cannot sufficiently praise the grace of their attitudes, the variety of their expression, the delicacy of the modelling and the elegance of the drapery; their exquisite style recalls the marbles of the temple of Wingless Victory, and they are undoubtedly the work of a master of the Attic school. Amongst them we should mention a fine equestrian statue, a remarkable head of a dying Amazon, an admirable torso of Victory, and two figures of Nereids on horseback which were probably placed at the corners of the pediment. Lastly, three statues of winged Victories in Parian marble crowned the summit of the building, which by the boldness of their pose, their exquisite refinement, and the impetuous motion of their drapery, strikingly recall the Victory of Paeonius found at Olympia and the Nereids which decorate the monument of Xanthos. They should perhaps be attributed to some sculptor of the Parian school, of which one master, Thrasymedes, worked at Epidaurus, and whose chief, Scopas, seems to have been inspired more than once by the tradition and manner of Paeonius.[*]

Many other statues have been found at Epidaurus

---

[*] These figures of Amazons and Nereids were made after models by Timotheus. See Prof. Gardner, *New Chapters in Greek History* p. 360.—E.R.P.

which bear witness to the length of time through which
the prosperity of the sanctuary endured. The greater
number represent Asclepius, and some of them seem
to reproduce a famous image of the god, perhaps the
seated statue which was shown within the temple,
while others are figures of divinities associated with
his cultus,—the triple Hecate, Hygieia, Aphrodite and
especially Athena. Most of these works, however,
date from a late period, and do not merit much attention. An Aphrodite however of the first century B.C.,
offers a fine and interesting replica of the Venus
Genetrix.

Before we end we may ask of what use these temples
were, and what we are to think of the part they
played in ancient times. Much evil has been said of
the Asclepieia of late: the priests of the god have
been considered notorious charlatans who impudently
traded on the credulity of the simple and ignorant.
Some have thought that they discovered among the
ancients themselves a deep dislike to these trickeries.
Undoubtedly the miraculous cures of Asclepius early
aroused some scepticism, as the inscriptions of
Epidaurus prove, but even this was not without
benefit for the sanctuary. When faith grew cool, and
the patients began to doubt the miraculous power of
Asclepius, the priests, in order to preserve the reputation of the temple, felt the necessity of applying to the
sufferers the resources of scientific medicine. This is
shown by a curious inscription from Epidaurus,[*] where
the treatment of dyspepsia is set out at length. The
prescription is very complicated, and I should not like

Ephemeris, 1885, p. 220.

to guarantee its efficacy, but these are the principal directions: never to give way to anger, to submit to a special diet of bread and cheese, of parsley and lettuce, of lemon boiled in water, and milk with honey in it, to run in the gymnasium, to swing on the upper walk of the sanctuary, to rub the body with sand, to walk barefoot before bathing, to take a warm bath with wine in it, to bathe and rub oneself, and to give a drachm to the bathing-man, to rub oneself with salt and mustard, to gargle the uvula and tonsils with cold water, finally, and this is all-important, to sacrifice to Asclepius, and not to forget to pay one's fees when going away. The treatment was to be tried for nine days, and the effect would be excellent.*

The value of the institution cannot be denied. The temples of Asclepius were, in truth, vast benevolent institutions or hospitals, doing their work under the eye of the god, with the encouragement and aid of the state. There the rich found useful guidance, the poor a refuge, and physcians themselves salutary inspiration.†
No doubt the institution degenerated in time; the temple of medicine often became a refuge for every kind of superstition, but it is none the less true that medicine found its earliest cradle there, and that for a long time the cultus of the physician-god was one of the most popular in Greece.

* Translations of the text of this inscription in Reinach, *Chroniques d'Orient*, pp. 96-7.
† Girard, p. 126.

CHAPTER X.

THE EXCAVATIONS AT TANAGRA (1870–1889).

BOOKS OF REFERENCE:—
    Rayet, "Les Figurines de Tanagra au Louvre" (*Études d'Archéologie et d'Art*, pp. 275-324).
    ,, "L'Art Grec au Trocadéro" (*Ibid.*, 364-79).
    Heuzey, "Recherches sur les Figures de Femmes Voilées" (*Mon. Grecs*, 1873).
    ,, "Recherches sur un Groupe de Praxitèle," etc. (*Gaz. des Beaux-Arts*, 1875).
    ,, "Nouvelles Recherches sur les Terres-cuites Grecques" (*Ibid.*, 1876).
    ,, "Les Figurines Antiques du Musée du Louvre."
    Kekulé, "Thonfiguren aus Tanagra."
    Pottier, "Les Statuettes de Terre-cuite dans l'Antiquité."
    Pottier and Reinach, "La Nécropole de Myrina. Paris, 1886.
    Reports of Late Excavations in *Deltion Archaiologikon*.
    Martha, "Catalogue des Figurines de Terre-Cuite du Musée de la Société Archéologique d'Athènes."
    Haussoullier, "Quomodo Sepulcra Tanagraei Decoraverint." Paris, 1884.
    Pottier, "Quam ob Causam Graeci in Sepulcris Figlina Sigilla deposuerint."

AMONGST all the provinces of modern Greece, Boeotia is certainly one of the least attractive and the least visited. Travelling there is difficult, the inns are full of discomforts, and hospitality anything but ready; while the scenery is rarely grand, and the ruins never striking; in addition to all this moreover the very name of the country, in spite of the great memories of Pindar and Hesiod, of Epaminondas and of Plutarch, awakes an indescribable notion of heaviness, clumsiness and hopeless stupidity. From ancient times Boeotia has had a most unenviable reputation : the Athenians, like good neighbours, were never weary of jesting upon the stupidity and awkwardness and ignorance of the Thebans ; their slanderers were always ready to comfuse Boeotian with imbecile ; and, indeed, when we cast up the intellectual balance-sheet of Boeotia and see that, with two or three exceptions, it has contributed nothing to Greek literature, and has never produced a great sculptor nor a great painter, we are inclined to think that its reputation was not undeserved. We should still be of this opinion if, during recent years, excavations had not chanced to reveal to us, in this ill-famed region, wonders of taste and refinement, of fancy and grace ; we mean those exquisite terra-cotta statuettes generally known as Tanagra figurines, from the name of the town where the greater number of them were found, but many charming examples of which have been discovered in all parts of Boeotia, at Thisbe, Thespiae, Thebes, Abae, and at other places as well, which amply avenge the Boeotians upon the contempt of antiquity.

I.

If from the heights of Parnes we turn our eyes towards the north, we see at its foot a very long and fairly broad depression running to the east towards the straits of Euboea ; it is the valley of the Vourienis, the Asopus of the ancients, a stream which possesses this peculiarity amongst others, that its channel only runs dry for a very small part of the year. In the centre of this valley, at the confluence of the Vourienis and one of its affluents, lie the scanty remains of the ancient city of Tanagra, quite close to the modern village of Skimatari. If the saying is true, Happy is the people which has no history, the inhabitants of Tanagra must have enjoyed perfect felicity, for we know in fact almost nothing of the history of their town, except that it enjoyed a certain strategical importance from its position where several important roads met, and that its possession was keenly disputed in the fifth century B.C. between Thebes and Athens. In later times, when the battle of Chaeronaea had overthrown the power of both Thebes and Athens, Tanagra, which was already a flourishing town, saw its importance still further increased; and after the destruction of Thebes by Alexander had rid it of a formidable rival, it became for some centuries the most populous city of Boeotia. A traveller of the second century B.C. describes it in these terms* "The city is steep and lofty, white in appearance and clayey. The interiors of the houses are elegant, and decorated with encaustic paintings." Life there, it seems, was easy and agreeable, the wine good, the people courteous, hospitable and charitable, the cock-fights famous throughout Greece; so that altogether Tanagra was an earthly paradise. In addition to this its women were beautiful, "the most comely and graceful in all Greece," says an ancient writer, "from their shape, their bearing, and the rhythm of their movements." According to the same traveller, there was nothing Boeotian in their conversation, and their

* Dicaearchus, *Frag. Hist. Graec. ed.* Müller *Die.*, fr. A. S. 9. 10.

voices were full of charm; it is therefore easy to understand why strangers found Tanagra so agreeable, and also why a poet of the period said, "Make friends of the Boeotians, and do not shun their women; for the men are worthy fellows and the women are delightful."

Nevertheless, in spite of the memory of their charming manners, in spite of the love of the beautiful which seems to have been widely diffused in the town, in spite of the splendour of its temples adorned by famous works of art from the hand of Calamis, and of the fame of its poetess Corinna who more than once was victorious over Pindar himself, the name of Tanagra would be completely forgotten if some astonishing discoveries, made about twenty years ago, had not drawn attention to this obscure district of Greece. For a long time the peasants of the neighbouring villages, when tilling their ground, had come across ancient tombs full of vases or statuettes; the name Skimatari, village of figurines, no doubt arose from this, but the objects themselves, found in small numbers, passed through so many hands before they reached a final resting-place that all precise indications of their origin were lost. It was not until 1870 that the explorations were pushed on more actively. A Greek from Corfu, Giorgios Anyphantis, better known under his nickname of Barba-Jorghi (old George), was just at this time engaged in secretly exploring the burial-ground of Thespiae; he heard a report of the discoveries which had been accidentally made at Tanagra, and established himself in the village, where, thanks to the experience he had had in work of this kind, he soon made the most splendid discoveries. Until his arrival the tombs explored had belonged almost exclusively to a very early period, and their contents were not very valuable; he was fortunate enough to find tombs of a later date, more richly provided, and containing objects which fetched a much higher price, so that in a short time he made a very handsome profit. Encouraged by his example, the peasants of the village left their farms and also began to excavate, so that the windows of the Athenian dealers in antiquities soon contained a large number of exceedingly lovely terra-cotta figurines. The prodigious success which these figures met with in Europe, and the rapid increase in price which followed in consequence, brought sudden riches to the people of Skimatari and redoubled their zeal. All the land of the village was dug up and turned over in every direction, and as the Greek government made no effort to organise regular excavations, and the Archæological Society of Athens did not condescend to take any interest in these delightful discoveries, the burial-ground of Tanagra was literally plundered by ignorant men whose chief anxiety was to make some lucrative finds. When at last the authorities bethought themselves that the excavations were unauthorised, and the Archæological Society realised that the finest statuettes had been taken out of the country, it was too late. It was useless to send a representative of the Society in 1873, supported by a detachment of soldiers, to put an end to clandestine researches with a strong hand; all that was of value had long since crossed the frontier, and nothing remained but to collect the refuse.

Three times the Archæological Society returned to the charge: in 1873, in 1875, and in 1876 it ordered excavations to be made in the burial-ground; but as its chief anxiety was to fill its empty cases, it cared little for making scientific explorations, and thus in spite of the great number of tombs which have been opened,— not less than eight or ten thousand—the excavations at Tanagra have furnished us with thousands of exquisite figures, but with no exact and certain information regarding the arrangement of this important cemetery.

Nevertheless it is possible, by means of inquiries made upon the spot and information furnished by some better instructed visitors who chanced to be present at some of the excavations, to form some notion of the circumstances in which the contents of the tombs in the burial-ground of Tanagra were found; and fortunately new explorations set on foot by the Archæological Society in 1888, and carried on this time on a really scientific plan, have recently supplied what was lacking in our imperfect knowledge. On the other hand, the excavations in the cemetery of Myrina, near Smyrna, where during three successive years (1880-82) under the direction of the French School at Athens, nearly five thousand graves were opened, afforded the most valuable information as to the arrangement of the ancient burial-ground, and brought to light tombs very similar to those of Tanagra. If we combine these different sources of information we shall understand with sufficient exactness how the Greeks buried their dead.

All along the roads which in ancient times stretched from Tanagra out into the country there ran long lines of tombs, grouped into three large cemeteries; naturally the tombs which they contained were not all of the same period nor of the same shape. Sometimes the grave is a simple quadrangular trench, a large hole dug in the ground, without any lining to protect the corpse; sometimes the bones of the dead are contained in sarcophagi of terra-cotta or stone placed in the earth; sometimes—and this is the commonest kind—the grave cut in the tufa is protected by a lining of bricks or by slabs of stone. "The great anxiety of the ancients was to give their dead a strong and inviolable retreat in order to ensure their repose, and at the same time to protect from desecration the objects of value often placed in the tomb. It was for this purpose that they endeavoured to make the walls of the tomb indestructible and to close it as completely as possible, either by filling the grave with a thick layer of earth or by covering it with heavy slabs of stone forming a kind of lid."* At the bottom of the grave the corpse was laid, with the head turned to the east or to the north—in this matter there was no unvarying rule—and all around it, mixed up with the bones and the earth in the tombs, were objects of all kinds, buried with the corpse and forming its funeral equipment. There were objects which the dead had used in daily life,—strigils and mirrors, boxes for paints and perfumes, ornaments and children's toys; there were vessels, too, designed to hold their food and drink, dishes of earthenware and bronze, cups and platters, bottles and lamps; there were also coins, and lastly figurines of terra-cotta. Naturally we do not find in all the tombs opened a complete set of the objects forming the equipment of the grave;

* Pottier and Reinach, *La Nécropole de Myrina*, p. 63.

statuettes, in particular, were a luxury which only the rich could offer to their dead. Very often a common unpainted vase placed at the head of the corpse and a few other worthless objects, were the whole garniture of the grave. It is only in a few tombs that we meet with the terra-cotta figures, which are sometimes entire but more often broken into a hundred pieces. As a rule the same tomb contains a considerable number of them; sometimes fifteen or twenty have been found in a single grave, and sometimes as many as fifty. They are not found however, as we might expect, arranged in good order as if on the shelves of a cabinet, but they have been thrown in carelessly, wherever there was space, between the wall of the grave and the corpse, generally on either side of the head or at its feet; many of them have been broken in the fall; many were intentionally dashed to pieces, to remove all temptation from those who might despoil the grave; while many, thrown into the tomb during the burial, have suffered still more. Consequently the number of figurines found entire is very small, and a too perfect state of preservation is almost calculated to throw a doubt on the genuineness of the figure. The scattered fragments of these fragile relics can fortunately be put together without difficulty; this is even, we may add, a favourite employment of the forgers, who, by dint of restoring and repairing and arranging, make up hundreds of false figurines for the benefit of unsuspecting purchasers.

Amongst the objects laid in the grave these statuettes of terra-cotta form by far the most interesting class. They constitute a little world by themselves of infinite variety, in which we find every style, every fashion, and every period; figurines of men and women, statuettes of divinities and spirits, as well as grotesque and indecent ones; jointed figures like puppets, and hollow figures with a stone inside like rattles; animals of every kind, statuettes of every degree of merit, rudimentary or exquisite, coarse or finished, all differing from one another according to their circumstances and their date. In very ancient tombs terra-cottas are rare, and in their place we usually find fine vases painted in the oriental style; it has been generally noticed that the tombs in which we find few figurines are rich in vases, whilst those in which statuettes are abundant contain hardly any pottery, as though these two forms of funeral adornment had succeeded one another. Nevertheless, as early as the seventh and sixth centuries B.C., the tombs contain very quaint figures, singularly naïve in conception and exceedingly coarse in execution, which the peasants, struck by the resemblance they offer, by their long garments in straight folds and the high caps upon their heads, to the dress of Greek monks, call παπάδες or priests, and which undoubtedly possess a religious character. They recall the ξόανα of early Greece, which were for so long the objects of popular veneration; their form is shapeless, their garments are only indicated by a number of coloured lines, their legs are not separated, and their arms are only stumps, while the profile is sometimes scarcely traceable, and is sometimes accentuated until it becomes like the beak of a bird. They are sometimes represented as standing, and sometimes as sitting on a high-backed seat, they are of no artistic value, and are only interesting from their antiquity.

In the tombs of later date, in which we can discern the approaching perfection of Greek art, terra-cottas are more numerous. Instead of these shapeless figures, we meet with works whose beauty is already striking; figures of divinities, easily recognisable from their attributes, busts of noble style which were placed on the walls of the tomb, and whose figures, cut off at the waist, seemed in a manner to spring from the ground, and symbolised the divine ascension of the chthonic deities. The greater number of these busts represent a veiled woman of majestic expression, whose gentle gravity and rather severe grace evidently denote a divinity, probably Demeter, the great goddess of Eleusis.

Most of the tombs, however, contain a greater variety of figurines. By the side of the statuettes of divinities, whose majesty, moreover, seems to grow more human, we meet with a great number of familiar scenes from daily life. There are children playing with their favourite animals, or carrying in their hands their balls and their knuckle-bones; there are ephebi equipped for the gymnasium or the chase, elegantly dressed women playing or walking, fashionable ladies and *demi-mondaines*, the thoughtless and the melancholy, the laughing and the sad, whose incognito it is not always easy to pierce. Then there are genre scenes, statuettes of animals, and jointed figures ; and all these fragile wonders, in which an exquisite feeling for form is united with a profound understanding of life, are admirable in their animation, their fancy, and their brightness. In spite of certain differences in style and workmanship they all spring from the same inspiration, and belong to the same period of art, no less brilliant than ephemeral, which, however, suffices to ensure the glory of Bœotian workshops.

Unfortunately, it is very difficult, in the absence of precise information drawn from coins and inscriptions, to determine the exact date of the Tanagra figurines; this is the inconvenient result of the clandestine excavations and the frauds which accompanied them. The greater number of the statuettes, however, must, it would seem, be attributed to the fourth century. Just at this time, as Rayet says, a remarkable transition was taking place in Greek art: it was emancipating itself from tradition, and freeing itself from the religious austerity of the preceding age, and seeking everywhere the charm of individuality and life. Praxiteles and Scopas brought the gods of Phidias to the level of humanity, Lysippus devoted himself to portraits, Apelles worked from the living model; in all directions, even in literature as well, we find a new taste for realism. This is precisely the characteristic of the Tanagra figurines; the workmen who modelled the statuettes—coroplasts, as they were called—went boldly forward on the new road which art was treading in the fourth century. The clay of which they made their little figures lent itself better than bronze or marble to the inventions of an art attracted by all that was natural and familiar, and masterpieces of the most charming and ingenious grace and fancy were produced by an inspiration, thoroughly popular in its nature.

It is on this account that the statuettes deserve to be studied with the keenest interest. They reveal an art which is quite new to us, with whose methods we were scarcely acquainted, and whose perfection we did not

suspect. They introduce us to the everyday life of
ancient Greece, its customs and pleasures and habits
and dress, they bring the contemporaries of Alexander
to life before our eyes, and, above all, they are charming
in themselves. Less than this would have made
Tanagra the fashion ; and, in fact, all the museums of
Europe contended for these exquisite and fragile figures.
Since the year 1875 the Louvre has been so fortunate
as to gather together a number of examples, almost
all of which are of value; private collections at the
same time, particularly those of MM. Lécuyer, de
Clercq, Dutuit, Bellon, Gréau, Rayet, were enriched
by charming statuettes, and, in spite of the acquisitions
made since then by the British Museum, and the
purchase of the splendid Sabouroff collection by the
Hermitage Museum at St. Petersburg, France has
become the possessor of the greater part of the spoils
of Tanagra. The exhibition at the Trocadéro in 1878
revealed the exquisite art of these Boeotian coroplasts
to the general public and set a seal upon their fame,
so that now even laymen are acquainted with these
masterpieces which the soil of Boeotia has restored to
us by thousands, and pride themselves on appreciating
their graceful ingenuity, their lively fancy, and their
unexpectedly modern spirit.

II.

" The Tanagra statuettes vary considerably in size :
the largest are as much as fifteen inches in height,
while the smallest only measure between two and three
inches ; but the greater number reach a height of about

eight inches when seated, of from five to seven when kneeling, and of eight to ten when standing. The appearance of all, however, is the same, and they are all made in the same way. Generally it is only the front which is carefully modelled; the back presents simply a rounded surface, in which contours and draperies are represented in a very sketchy way, and in the middle of which is a rectangular hole, sometimes of considerable size—the vent-hole, made in the figurine when the clay is still soft, which served to facilitate evaporation during the baking. Almost all are placed on a very shallow plinth added beneath the statuette before it was fired; some stand on a lofty pedestal which forms part of the figurine." * Most of them were made in a mould, but the figure thus formed was afterwards touched up with modelling tools, and completed and corrected so that each figurine received an individual stamp under the hand of the coroplast, and thus the mechanical reproduction became a work of art. It is on this account that the Tanagra figures deserve a place in the history of art, and that the way in which they were made deserves to be studied with some attention.

We shall not dwell upon the very simple technique adopted in the manufacture of early examples and those of little value, such as cheap idols, dolls, and children's toys; all these cheap goods were made by hand, and naturally in a very coarse way. The use of the mould, which made the work at once easier and more certain, and by leaving the inside of the figure hollow, also made the statuettes much lighter, soon

* Rayet, *Etudes d'Archéologie et d'Art*, pp. 287-88.

took the place of this primitive method, the only one with which the simple workmen were acquainted who fashioned the first quaint figurines stamped out of a thin cake of terra-cotta. It is easy to imagine how the coroplast set to work; with a first layer of fine, well-kneaded clay he took the impression of the mould in terra-cotta by pressing the clay with his finger into all the cavities of the mould; then to the first layer he applied a second, and so on, until he obtained the required thickness. In many statuettes we can still trace the marks of the workman's fingers on the inner side of the figurine. The mould was usually in two parts, one of which represented the front and the other the back of the figure; the clay was pressed into each of the two halves, and then the moulds were brought together, and the two parts of the statuette made to adhere on the inside with a little clay. The head, arms, and feet, were also, as a rule, made separately. Naturally, on its removal from the mould, often worn by long use, the cast presented a rather rude appearance. Here began the true function of the coroplast: he set to work afresh on this rough model of a statuette, worked it up, and finished it off; he emphasized the somewhat lax modelling, he hollowed the folds of the drapery with his graving-tool, rendered the head more expressive with a touch of his thumb, and delicately threw into relief the dainty edifice of the hair, thus impressing on the work the mark of his individuality. Nay, more: as the prominent parts were generally added later—as the head, feet, hands, and accessories of all kinds, hats and fans, balls, wreaths, wings and the rest, were made separately—the figurine had really to be dressed, and the coroplast could adorn it to please his fancy. He chose one or another of these little heads, attached its long neck to the top of the statuette, and fixed it with a touch of his finger; then he bent or raised it, and bestowed upon it according to his fancy a dreamy or coquettish, a sad or gay expression, thus giving quite a different aspect to two casts from the same mould. He would put a garland of flowers into the hands of one, and give a fan or a mirror to another, and his fancy combined the accessories of all kinds which were at his disposal with astonishing fertility of imagination.

If we wish to see some examples of the transformations, often very witty and amusing, which the coroplast's caprice made the different copies from the same mould undergo, a figurine from Athens will show us a piquant combination. The statuette as it left the mould represented a young girl picking flowers; some fancy crossed the mind of the modeller, and without stopping to correct the feminine contour of the bust, he worked it over so as to represent the roughness of a shaggy skin; between the shoulders he planted a grinning head, with pointed beard and long ears,—and behold the graceful girl transformed into a satyr! On the body of a young man, dressed in a short cloak, the coroplast would put different heads in turn, sometimes he would make him wear a cap and sometimes a pointed hat. One day chance threw a winged cap into his hands, and the ephebus became a Hermes. Further on we shall find two young girls resting on a seat without a back in a charming attitude of reverie; both are alike, and no doubt they came from the same mould, but to

one the artist has given a dainty little head, the hair
dressed to perfection, while he has bestowed upon the
other a face wrapped in a veil and wearing a sorrowful
expression, thus rendering the two sisters absolutely
unlike. When the mould represents a group, the
variety is still greater, for an alteration in the slightest
detail gives it a new aspect. The motive which
represents a pair of lovers has been reproduced to
infinity in the coroplasts' workshops; and still they
were able to present this uniform type in a thousand
different ways; by the change of a head or an attribute,
they represented in turn Aphrodite and Eros, Aphrodite
and Adonis, Dionysus and Ariadne, Dionysus and
Methe. Lastly (and this is a new source of piquant
variety), when a mould was worn out a new impression
could be taken from a particularly successful copy,
and by this double contraction there was produced a
statuette of the same type but much smaller, and in
this way a series could be formed of identical representations gradually decreasing in size. Thus with a
very small number of moulds the coroplast produced a
display of marvellous originality and novelty, and we
cannot sufficiently admire the fertility of imagination,
the changeful fancy, and the prodigious dexterity with
which these unknown artists succeeded, by skilful retouching and ingenious combinations, in giving infinite
variety to a small number of types.

The toilette of the figurine was, however, not yet at
an end. When it had been retouched, it was allowed
to dry; it was then fired, and returned to the workman's
hands in order to be painted. It was first dipped in a
bath of lime or chalk, which formed a creamy white

surface which took the colours better than the porous clay, and then it was painted from head to foot. The drapery was generally coloured blue, red, or pink; but besides these tints, black, yellow, green, and a light violet were often used. Jewels and ornaments, diadems, bracelets, and earrings were gilded, the hair was painted a beautiful reddish brown, recalling the auburn tresses of which the Boeotian women were so proud, the lips were red, the pupil of the eye pale blue, while the black eyebrows, very much prolonged, recall the fact that antimony was known in ancient times; the cheeks, lastly, were given a pale pink hue. Sometimes the colours were fixed by being fired a second time with a gentle heat, but this is the exception rather than the rule, and consequently many of the figurines have lost their original brightness through their long burial, though many statuettes have been found both at Tanagra and at Myrina which have retained an incomparable freshness of colouring.

It is difficult to realise the astonishing favour which the Tanagra figurines found throughout the ancient world; in Asia Minor, Italy and Sicily, and from Cyrene up to the Cimmerian Bosphorus, wherever terra-cotta statuettes were made, the products of Boeotian workshops were eagerly sought after and imitated. In all these workshops, moreover, the methods of manufacture were nearly the same. The art of the coroplast is essentially an industrial art—one which has its fixed rules, and in which mechanical labour has a large share. It is true that the originality of the artist makes itself felt in ingenious and unforeseen combinations of accessories, in the skill and intelligence displayed in retouching and in the tasteful harmony of colours, but, before all, the workman is practising a trade by which he has to live. This must never be lost sight of in endeavouring to interpret the meaning of these graceful statuettes, for where we endeavour to find the traces of premeditation and profound thought, there is often nothing more than a fanciful caprice, the whim of an artist desirous of showing with what ingenuity he can vary a single type, and how tastefully he can group the most heterogeneous elements. "Suspended between the ideal and the real world," M. Heuzey has gracefully said, "many of these figures remain in an uncertainty which forms part of their grace. They are fragile and delicate things, which science must not touch with too heavy a hand lest it should crush them." Would that the commentators had been guided by these words of wisdom, and that the author himself had not sometimes forgotten them a little.

III.

During the twenty years, in fact, that we have been acquainted with these Tanagra figurines, the question has often been put, What do these graceful and dainty little figures represent, with their piquant air, their gait now rapid and agile, now indolent and languishing, and their exquisitely graceful attire? And upon this difficult question there is discord in the camp of the archæologists. On the one hand, M. Heuzey, a champion of delicate taste, undoubted learning, and marvellous ingenuity, endeavours to show, with astonishing fertility of argument and remarkable skill, that these figurines so delicate and *spirituelles*

have a religious and symbolical sense, and that under
their mundane appearance are concealed the great and
mysterious divinities of the lower world. On the other
hand, a whole school of archæologists, adopting a
simpler and more ordinary explanation, seek for representations of daily life in these graceful statuettes, and
will see nothing in them but genre subjects. The battle
between the two systems has lasted long, and among
the most ardent of the combatants, among those who
have fought most brilliantly in this sometimes epic conflict, we cannot forbear to mention Oliver Rayet, so
early taken from the study to which he had rendered
and might still have rendered such great services. As
always happens in battles of this kind, there are neither
victors nor vanquished and both parties have encamped
on the field, but now, when the heat of conflict is over,
it is perhaps possible to find a middle way between the
contradictory opinions which will doubtless be nearer
the truth than either.

We must certainly admit as an indisputable fact
the point which serves as a basis for all M. Heuzey's
demonstrations: that is, that in tombs of comparatively
early date, from the period of archaic art to that which
borders upon the period of perfection, the figurines
always represent divinities, and are, in fact, idols belonging to the grave. These figurines, stamped from a
cake of terra-cotta, with their calm and solemn attitude,
their strange and lofty head-dress, undoubtedly possess
a religious character, these busts, stamped in low relief,
in which definite attributes enable us to recognise
Demeter, Persephone, and Dionysus, these figures cut
off at the waist, whose expression is so noble and whose

style is so perfect, evidently represent the powerful deities of the lower world, leaving their subterranean home to ascend among the gods; and I need not refer to that significant gesture, which with one hand brings one of the folds of the veil over the breast, a delicate and modest transformation of the Oriental type, which by the two hands pressed upon the breast naïvely represents the goddess-mother who sustains life in the world and among human beings.

It becomes, however, more difficult to follow M. Heuzey, when from these solemn figures he passes to the slender and graceful figurines of the fourth century, whose expressive faces, and whose attitude and attire, do not immediately suggest types borrowed from religious and heroic myths. Undoubtedly we may admit that races do not in the course of a few years change their traditional ideas concerning the life beyond the grave, and yet it is very difficult to recognise the mysterious divinity which conceals itself in these nameless beauties of such mundane grace. We cannot fail to admire the extreme ingenuity and marvellous flexibility with which the slightest caprices and the most fugitive thoughts of the ancient modeller are interpreted in favour of the theory, and it is interesting to see how, under M. Heuzey's pen, everything conspires to make us recognise the great goddess Demeter in this youthful female figure with its slow uncertain steps and its somewhat sad and melancholy expression; how everything becomes significant in this charming figurine, even to the blue colour of the veil, the hue of sadness, to the broad, pointed hat which befits the wandering goddess traversing the world in search of her daughter, even indeed to the absence of characteristic attributes and the effacement of the divine type, by which the artist wished to indicate the mystery in which the chthonic deities loved to shroud themselves. In this way, everything in these little folks of Tanagra can easily be explained. If the veiled figure is accompanied by a younger companion, clothed in bright colours and with uncovered head, we must see in this group the happy reunion of Demeter and Cora, while if we are called upon to identify this young girl kneeling on the ground and resting one hand on the earth, we think at once of Cora picking flowers in the meadow of Nysa, or at least of some immortalised being playing with her dice, continuing the pleasures of life beyond the grave, and symbolising with profoundly philosophical intent " the contrast between the full bloom of womanly beauty and a tragic fate." If we betray some astonishment at finding these profound thoughts hidden beneath such slight and trifling appearances, and wonder that the ancient deities are content with such an easy grace, such a familiar fashion of representing the gods, M. Heuzey explains this profound change by the transformation which art and religious feeling underwent in Greece in the fourth century under the influence of Praxiteles. These softened types are none the less easily recognisable ; they represent divinities now as formerly, —Demeter and Cora, Dionysus and Aphrodite, who in the fourth century enters the cycle of the goddesses of the dead ; around them cluster a whole train of beings of a nature more familiar, but not less divine, nymphs and spirits of the lower world bearing the poetic emblems of Elysian bliss, followers of Aphrodite, bacchantes forming part of the thiasos of Dionysus, veiled attendants on Demeter, a whole new and charming cycle closely connected with the beliefs and the funeral rites of Greece, and forming an utter contrast assuredly to those genre subjects created indifferently, apart from religious thought, by the caprice of the modellers.

A superficial analysis can give no idea of the unrivalled skill with which M. Heuzey supports his thesis, and the refined grace of the style in which he clothes his supple dialectics. Nevertheless, in spite of the seductive charm of his ingenious essays, we cannot deny that there are remarkable differences in style and inspiration between the idols of primitive times and the figurines of the fourth century, and that it is almost irreverent to recognise divinities in the fashionable ladies who are quite absorbed in arranging their draperies gracefully, in looking behind them to see the effect of their trains, and in trifling with their fans and mirrors, when they are not casting bewitching glances around them. A large number of statuettes, too, are very much out of harmony with the lofty meaning M. Heuzey ascribes to them. The women prosaically engaged in putting their bread into the oven, the good barber of Tanagra just about to curl his customer's hair, the hawker crying his wares, the child writing his copy under the master's eye, and the other child fighting a strange battle with a goose as big as himself, the cook getting his luncheon ready, and many others—are all these religious subjects? have they the slightest connection with burial rites? Have these grotesque figurines, these grinning and distorted faces of comic and tragic actors, these often obscene caricatures, any relation to the deity? It may certainly be said that the object of these statuettes was to amuse the dead man and thus to overcome his ill-will against the living, or that they are part of the train of Bacchus, of his escort of jesters and actors, but, as a matter of fact, if we look with an unprejudiced eye on a fairly numerous collection of these figures, they produce upon us a very definite impression. Certain figures are unquestionably representations of divinities, others as undoubtedly bear an entirely human and familiar character, but between the mythological subjects and those inspired by the scenes of daily life there are a multitude of doubtful statuettes whose meaning it is difficult to discover. It is round these fair figures of uncertain origin that the battle rages, according as we bring into prominence their religious signification or dwell upon their realistic and picturesque side, and it is to themselves that we must look for the solution of the question. We may perhaps find that it is not impossible to reconcile two theories each of which contains a part of the truth.

The art of the coroplasts was above all industrial and popular—"a sort of trade in cheap images," in which the maker worked at great speed, and concealed the hastiness and sketchiness of his work under an air of pleasing negligence. Consequently we must not look for any very profound aim or very philosophical intention in this rather hurried method of production, which repeats by dozens the impressions from the same mould. The sculptor who spends whole days in hewing out of the marble an original figure may be able to make every detail in this patient work of genius contribute to the realisation of his thought, but the coroplast, as

M. Martha says, is an artisan who makes a figure with somewhat mechanical indifference, and who is in haste to have done with it that he may pass on to another piece of work. Moreover, as he works for a living, the little figures exposed for sale at his stall must take the fancy of purchasers, and therefore they must be suited to the prevailing taste. If the fashion is in favour of religious figures, he will manufacture idols for the grave; when art emancipates itself, and under the influence of Praxiteles the gods of Olympus assume a more human aspect, the coroplast transforms the primitive types and arrays the old divinities of Hellenism in more modern dress. On the other hand, it is inevitable that subjects which are originally religious should in time lose their primitive meaning. By dint of repeating and reproducing the original composition according to the traditions of the craft, the meaning of the myth is gradually lost; the study of nature, to which the artist of necessity devotes himself, introduces variations, and the sacred type becomes nothing more than a mechanical reproduction of a workman's pattern in which the desire for a beautiful and picturesque effect has gained the upper hand. At this moment it would be very difficult to say whether the coroplast has forgotten the religious thought expressed by the primitive type, or whether he has intentionally rejuvenated the sacred model by a piquant blending of symbolism and realism; whether he has simply, to adopt an ingenious comparison, transposed a grand and noble air into an easier and more familiar key, or whether he is playing a new air unconsciously inspired by ancient themes. The transition from a religious to a genre subject has become so easy that the step is often taken, but whilst the symbolical meaning has disappeared, the plastic composition retains its essential features, and an obscure connecting-link often binds together two groups which seem to support two contradictory theories. Certainly we cannot seek for a symbolical or mystic meaning in every statuette, but on the other hand it is impossible to deny the persistent influence of religious traditions; the desire of rejuvenating ancient types by picturesque effects is capable of insensibly changing their character, but scenes of every-day life are often closely connected by some bond of origin with religious subjects.

Another and not less difficult question arises if we inquire for what reason figurines were placed in the tombs, and in order to reach a solution it may be worth while briefly to recall the conceptions which the Greeks entertained of the life beyond the grave. For them, as for all the other people of antiquity, life did not come to an abrupt close at death, but in the tomb where the body was imprisoned an obscure existence was maintained with all the needs and pleasures and desires of humanity. Even at a later time, when the Greeks pictured to themselves all the souls of the dead assembled in Hades, a subterranean region vaster than the tomb, their only conception of this future life was as a repetition of life on earth. It was therefore the duty of the living to supply food to the dead, who continued to exist within the tomb; and this is the reason why wine and cakes and milk were placed upon the grave, and also why, on certain anniversaries, funeral banquets were celebrated there, at which the shade of the dead

man was present though invisible. It was also the duty
of the living to see that in the solitude of the tomb the
departed were surrounded by the objects they had
cared for on earth, and therefore arms, gymnastic
appliances, mirrors, needles, boxes of paints and cases of
perfumes, were buried with them. They must not only
be provided with necessaries, but with superfluities as
well, they must take their friends and companions down
with them into the other world in order to recommence
their round of pleasures there; for this reason their
horses and dogs were buried with them, and in early
ages slaves and captive women were often sacrificed
upon the grave, that they might go down into Hades to
wait upon the departed, or to enliven his loneliness. In
later times when manners became less barbarous, these
cruel customs disappeared, and bloodless sacrifices,
prayers and music offered at the grave took the place
of these sanguinary rites; but still the idea remained
that the solitude of the dead man must be enlivened,
and the melancholy of his shade dispelled. To cheer
the departed in the depths of the tomb, and to protect
him against the dangers of that mysterious journey,
was the twofold desire by which the piety of the survivors was inspired. It was for this purpose that the
Egyptians placed statuettes in the tomb, to answer the
summons of the departed, to aid him in the cultivation
of the celestial fields, to form a devoted escort around
him, and to secure him immortality. The Assyrians,
from a similar motive, placed in the graves figurines
designed to avert the hostility of the chthonic powers,
and this too is the object of the sepulchral idols found
in ancient burial-grounds at Rhodes, which represented

the guardian divinities of the tomb and afforded escort and society for the departed. This is also undoubtedly the reason why the cemeteries of Tanagra and of Myrina are full of terra-cotta statuettes; but this question is still keenly disputed according as we look to one or the other of these two dominant ideas—the wish to protect the dead, and the wish to provide them with company in the grave.

M. Heuzey, who recognises divinities in all the figurines found in the tomb, easily arrives at a solution of the problem: it was to protect the departed on their dread journey that the piety of the survivors placed around them the images of their gods, and naturally the first place among these images was given to the mysterious powers of the lower world. This was the object of the archaic figures, of the sepulchral idols of noble style, of the fourth century figurines of more familiar aspect, and even of those grotesque or ridiculous figures whose presence causes some astonishment at first. Even this buffoonery, however, played a useful part: it was a protection against the evil spirits and malevolent influences which surrounded the departed; it provoked a laugh in their gloomy subterranean prison, and laughter was believed by the ancients to have a beneficent effect; it diverted the dead, and lifted the inauspicious gloom of the grave.

It cannot be denied that the very ancient figurines do in reality represent the guardian divinities of the dead. May there not, however, be some imprudence in attributing a strictly symbolical and sepulchral meaning to all these statuettes? Terra-cottas have been found in the greatest numbers in the burial-grounds, but is this equivalent to admitting that they are only found there? We shall very soon see the contrary. Among the objects found in the tombs, are there not many articles of domestic use or belonging to the toilette, pins and mirrors, collars and bracelets, to which no one has even thought of assigning an exclusively sepulchral meaning? Hence the other view, that the terra-cotta figurines are not the guardian deities of the deceased, but simply the articles he had cared for—the charming toys which during his life had adorned his home on earth, and which went with him into the grave in order to adorn his dwelling in the shades. Formerly they had formed the decoration of his domestic shrine or had been arranged in graceful lines on the walls of his house; now—as was very natural—they followed in death the master who loved them and took delight in them.

Some of these statuettes, notwithstanding, have such a clearly sepulchral character, that many, while denying that they represent divinities, are inclined to connect them closely with the cultus of the dead. In Rayet's opinion, their essential aim was to keep the dead man company in the solitude of the grave. Formerly, as we have seen, he was given real companions, sacrificed upon his tomb; in later times, these living victims were replaced by simple imitations, in the same way as real jewellery was replaced by gold leaf, precious stones by glass beads and gilded beads of terra-cotta and the food and flowers formerly offered to the departed by similar articles in terra-cotta which were not costly and would last longer.

It might well be supposed that the shade of the jewels might suffice for the shade of the dead, and the piety of the survivors suffered them to make the substitution without feeling any scruples or becoming conscious of any irreverence.* Even the gods of Hellas were not offended by this kind of subterfuge; the dead were not harder to please than the living, and in default of the actual flesh-and-blood companions formerly sacrificed upon their tombs, they willingly accepted these terra-cotta companions, who recalled the pleasures and the employments of earth, and brightened by their presence the unsubstantial life which the shade still preserved in the depths of the tomb.

In order to harmonise these diverse opinions, it will perhaps be sufficient to take the different periods carefully into account. In early days the chief object in popular belief was undoubtedly to secure divine protection for the dead at the moment when they crossed the threshold of that lower world, of which no Greek in primitive ages could think without terror; and it is for this reason that in archaic tombs we find only sepulchral idols. But can we expect from the fourth and third centuries, from the sceptical age of Alexander and his successors, the strict symbolism and profound faith of the time of the Persian wars? Not only the religious beliefs but the character of the people had undergone a great transformation; new and more winning deities were insensibly substituted, even in the cultus of the dead, for the austere figures of the goddesses of the lower world; the meaning of the old traditions grew dim or faded away. In early ages the articles which were to bear the dead man company

* Rayet *Études d'Archéologie et d'Art.* p. 323, gives two examples. Thuc. Bk. I. 134.

were carefully placed quite near the corpse, while very often in later graves we find them heaped up carelessly above and outside the tomb which was already closed. The traditional habit of throwing terra-cottas into the grave had been preserved, but the reason of the custom had been forgotten. The subject represented by the figurines mattered little then; once they had been gods, now they were men; the essential point was to render the departed the homage he claimed and to bring a pious offering to his ashes.

If, in fact, we try to ascertain from existing remains and literary sources what use was made of these terra-cotta statuettes in antiquity, we shall find that they were by no means always destined for the sepulchre. In the charming passage at the opening of the *Phaedrus* of Plato, Socrates is walking on the banks of the Ilissus, and at the waterside, at the foot of a plane-tree, he meets with a shrine consecrated to the Nymphs and decorated with figurines of clay. Modern excavations show that offerings of this kind were often laid at the foot of the altars; at the sanctuaries of Asclepius at Athens, of Demeter at Tegea, of Athena Cranaia at Elatea, at Olympia, at Corinth, at Cyprus, and in Southern Italy, great heaps of terra-cottas have been found near the temple, the *ex-votos* consecrated by the faithful. In other cases figurines were used for strictly domestic purposes, and served either to protect the house or to adorn the interior of the dwelling. In the excavations at Pompeii, they have been found by hundreds, arranged in niches at the foot of the Lares or interspersed with the small bronzes which decorated the gardens and the rooms. Lastly, they served as toys for children and even for grown-up people. At the time of the Saturnalia the Romans sent one another clay figures as presents, and on vases we more than once find children and girls playing with terra-cotta figurines. What is most remarkable is that all these statuettes, both those that were consecrated in the temple and those that decorated the houses, toys as well as religious offerings, represent exactly the same subjects as those we find in the cemeteries.

It would be very imprudent, when we venture on an interpretation of these delicate and charming toys, to attribute to the coroplast who made them any conception especially connected with the cultus of the dead. We undoubtedly find in the cemeteries certain subjects which are especially sepulchral, like the winged Sirens with loosened hair, or the Loves with drooping heads and pensive faces, who symbolise the mourning and the sorrow of the survivors, but we only meet with a few figures of this kind, and the remainder would be equally suited for a private house or for the treasury of a temple. Several of the figurines too, which were discovered in the tombs of Myrina, evidently belonged to the dead man during his lifetime. Some still bear upon their pedestal their owner's name; while others, like the group of thirteen figures of different heights, arranged symmetrically and forming a descending series starting from a central subject evidently formed a decorative *ensemble* in the dead man's house, to which he was particularly attached, and which on that account was laid in his tomb.

There is then, properly speaking, no necessary relation between the terra-cottas and the graves. The coroplast, who made the figures in his workshop, was only desirous of making them pretty and attractive, in order to draw customers to his shop and to make a good profit from the work of his hands. If any curious person were to ask him what divinity he wished to represent, and to what use he meant to put it, he would simply answer, as some curious inscriptions placed on certain accessories at Myrina show, that he had modelled an ephebus, a young man, almost as French workmen speak of their *bon homme*;* the purchaser would decide the destination of the figurine, according as he wished to make an offering to a divinity, to adorn his dwelling, or to do homage to some departed friend; it was the intention of the donor, and not the subject represented, which would assign the statuette to a particular purpose. In Attica, where pottery was the chief manufacture, it was natural to place vases in the tombs, while in Boeotia, where the charming art of the coroplast was flourishing, figurines were purchased to do homage to the dead.

It is now easy to arrive at our conclusion. According to the sepulchral beliefs of the ancients, the dead man not only needed to be accompanied by his guardian divinities on his journey into the other world, but after death he became a sacred being, who might prove mischievous or helpful to the living; if they ceased to offer him the funeral banquet, or neglected the rites of his tomb, his shade would come out of his grave to reproach them with their impious neglect, while on the other hand he would love and protect those who paid him honour, and this protection was always effectual. The survivors had, therefore, an interest in conciliating

---

* Pottier and Reinach, *Nécropole de Myrina*, p. 195.

the good will of the dead and in showing them by their pious offerings the respect and attention they exacted. It was for this reason that together with sepulchral divinities to afford him escort and protection, and figures of objects specially expressing the regret of the survivors, they would buy any statue, no matter what, in the coroplast's shop; for the intention alone gave value to the offering; it was for this reason too that we do not find such statues in all the tombs, for such costly homage was not within the reach of every purse. In all these cases, however, just as the coroplast was not bound by strict rule and was chiefly anxious to make, not orthodox statuettes but elegant figurines suited to the prevailing taste, so the donor was not concerned about the subject his offering represented; he simply sought a pleasing object which might be agreeable to the dead and draw down his protection upon the living. For this reason he gave him the escort of his god, he buried with him the ornaments he had cherished during his life on earth, and cheered his solitude with companions; by this pious homage he paid his debt to the shades, while at the same time he deserved their gratitude.

We must not, I think take too gloomy a view of antiquity, and ascribe to its sepulchral rites the deep and subtle meaning of a more austere piety. The Greek burial-ground must have borne a strong likeness, I think, to those great Turkish cemeteries, full of shade and verdure, where the graves, piled up in picturesque disorder, do not arouse any thought of death. In the great cemetery of Scutari, whose cypresses climb the sloping shores of the Bosphorus, the thought of death and of the gloom which surrounds it is absent,

the rays of the sun glance brightly through the black
pyramids of cypresses, the birds are singing in the sky,
and the changing colours of the dresses of the Turkish
ladies, whose favourite walk is in the cemeteries, contrast gaily with the whiteness of the tombs and the
dark green of the turf. It would seem to have been
much the same in ancient times; a visit to the burial-grounds gave the women, usually shut up in their
apartments, the opportunity of taking a walk, and also
of displaying their charms and winning admiration.
On Attic vases we often see opposite the figure of a
woman who has brought sepulchral offerings to the
grave, a young man, a passer-by, who has stopped,
and seems to be taking a warm but not altogether disinterested share in the grief which is displayed beneath
his eyes.[*] Death, too, for the quick and lively temperament of Southern races, has never acquired the
darkness and gloom which are associated with it in
other countries, and has never aroused in their minds
the deep and melancholy thoughts so often expressed
in the poetry of Northern races. This should never
be forgotten in studying the statuettes found in Greek
burying-grounds, lest we should throw too deep a veil
of sadness over these exquisite and fragile figures.

IV.

It is time for us now to forsake these controversies,
and to seek for enlightenment from the Tanagra figurines
on other points, and while we assign its share to
mythology, and recognise among these charming little

---

[*] Cartault, *Terres-cuites antiques de la collection C. Lécuyer*, and
Terres-cuites Grecques.

people the undeniable presence of divinities, and especially of the troops of graceful figures, Loves, Satyrs and Bacchantes that accompany them, we will rather devote our attention to the types derived from daily life. These Boeotian statuettes will tell us more of the daily life of the Greeks in the age of Alexander—of their games and pleasures and customs, and above all of their dress—than the study of many a folio volume and the contemplation of the sublimest monuments of statuary, nor is this the least valuable feature of the discoveries at Tanagra.

The Boeotian coroplasts did not assign a very prominent place in their collection to the male sex. It may have been that clay, the material used by the modellers, is better adapted to render the rather soft and indeterminate contours of women and children, or that in this somewhat hasty industrial art the figurines which admitted of more careless and fanciful execution, those to which a touch of the thumb would give a striking and ingenious turn, were given the preference over figures whose contours were more noble and severe. The art of the coroplasts was always readily satisfied with a hasty and somewhat sketchy style of execution, and was satisfied to substitute a pleasing neglect for exactness of work, and to replace by charming indefiniteness the clear and precise contours of older masters. It is equally certain, whatever the cause may be, that masculine life is only represented among these statuettes by two types, that of the child and the ephebus. The artists' imagination has no doubt diversified these two themes with marvellous fertility, but it has never tried to invent any novelty.

Scenes from childhood have a special charm for the coroplasts. The graves are full of laughing and refractory little ones, who busy themselves with charming *naïveté* and grace with their childish games. Sometimes seated on a square altar, scantily clad in a little shirt, the child holds in its hand the net containing his ball or the bag with his knuckle-bones, sometimes he is trundling a hoop or spinning a top, and at other times he has harnessed himself to a little carriage, or is gravely wearing a comic mask, or carrying a musical instrument. The greatest pleasure of these children is, however, to play with their favourite animals, we find them sitting astride geese, cocks, and rams, generally on very bad terms with their steeds, and struggling with them with all the strength of their little arms; again there are epic duels with cocks and geese, mighty battles in which the combatants, the one as eager as the other, are hotly disputing the possession of a bunch of grapes.

Then the years pass on, and the child becomes a youth. We find him again on his way to school, accompanied by the paedagogus, who is carrying his gymnastic equipment and has him in charge. By the side of the master we find his pupils, sometimes on horseback and sometimes dressed for the palaestra, with a large felt hat upon the head, and a lance or an oil-flask in their hand. In other groups they are playing their favourite games, and are very often engaged in cock-fighting, one of the most popular of diversions at Tanagra and throughout the whole of Greece, and one which sometimes led young men into the wildest excesses.

The female figures are far more numerous, and their execution is, as a rule, more careful. There is a great diversity among them, and their classification is often difficult—attitudes, faces, gestures, accessories, are infinitely varied. There are love-lorn maidens whose spindles are in front of them, but who are paying less attention to the hateful wool "which makes women grow old" than to the little Love perching upon their knees, while others, melancholy and deserted, are sitting on a rock, their heads drooping and shrouded in veils, their thoughts lost in profound despair or in vague dreams. Fashionable beauties come walking by with proud grace, or looking coquettishly around them and displaying, with apparent carelessness, the elegant details of their dress. These charming figures bring up before our eyes women's life in all its phases, from the still childish pleasures of the young girl to the amusements and coquetry of the woman, and afford us a thousand unexpected revelations as to the way in which Greek ladies attired themselves, the colour of their dress, and the infinite resources and refinements of their toilette, the study of which has so much attraction for us.

The young girls of ancient Greece were not very well-informed, they would have found it difficult to discuss music or painting or literature with any understanding of the subject, and they had not even that insight into everything which was expected of Frenchwomen in the seventeenth century. During the years of idleness which preceded marriage they receive very little instruction. "What did my wife know," asks an ancient writer, "when I married her? She was not fifteen then, and up to that time the chief care of those

who surrounded her had been to keep her eyes and
ears closed and not to excite her curiosity. Was it not
enough for her to know how to make a cloak with the
wool that was given her, and to distribute their tasks
to her maidens? As regards discretion, she had
been taught excellent principles." Spinning, weaving
and embroidery formed the chief part of a woman's
occupations and of the useful knowledge expected from
a young girl; still these employments were not very
absorbing, and left a good deal of time for amusement,
as we can plainly see from the terra-cotta figurines.
The education these girls received was not at all grave,
and no very deep or serious thoughts had prematurely
clouded their brows. We find them playing with
their pets, especially with their doves, upon whom
Greek girls bestowed the nicest care, sprinkling them
with perfumed water and giving each a different scent.
Sometimes they are at their games, playing at ball as
Nausicaa loved to do, and displaying, if they choose, not
only their dexterity but their graceful and charming
attitudes as well. There were many ways of playing
this favourite game; some would exercise their skill in
keeping several balls up at once, while others found a
pleasure in letting it fall into a fold of their dress, an
amusement in which it is very difficult to see the point,
it seems so simple. Knuckle-bones was also a favourite
pastime, either thrown into the air and caught on the
back of the hands, or thrown to a distance into a circle
or a hole made in the ground. Many of the figurines
represent girls playing this game, looking after the dice
as they fly out of the box, or picking them up from the
ground, either alone or sitting opposite their partners.

There were other amusements, too, the excitement of the swing and the joys of the *encotyle*, a game in which the loser had to carry the victor on her shoulders for a certain time, and finally the pleasures of the bath. Thus the long hours went by in the retirement of the women's apartments, and this careless existence lasted until marriage. "Timareta," says an epigram in the Anthology, "consecrated to Artemis before her marriage her musical instruments, the ball she loved so much, the net which confined her hair, her dolls and their dresses. O goddess, it was natural that, a young girl herself, she should offer these little ones to a young girl." This offering of dolls would seem very tardy to us, our young girls *fin de siècle* are more precocious and their amusements less simple.

Marriage came at last to set these caged birds free. The constitution of the family, however, in ancient times, and the obligation, often imposed by law, of marrying a woman with whom one was scarcely acquainted, did not leave much room for sentiment in marriage. Whilst the bridegroom looked for a dowry, the bride hoped for an opportunity of giving fuller satisfaction to her love of luxury and dress. In the fifth century the strict seclusion of the women's apartments placed some check upon her whims, but the freer life of the fourth century gave full rein to her caprices. Her toilette, into which fashion had introduced many refinements, then became the chief business of every woman; all wore fine and transparent materials of varied hues falling in harmonious folds, all adorned themselves with jewels, and dyed their hair and introduced into its arrangement the most complicated devices and audacious novelties, they darkened their eyes and painted their faces, and as men were the first to set this example, we may imagine that they did not lack eager imitators.

Nothing was more simple than the dress of a Greek woman. Its essential part was the talaric chiton, somewhat resembling the long, embroidered garment worn by Albanian women. In this dress the skirt and the bodice are undivided; sometimes it has short sleeves, sometimes it is open at the top, and buttons on the shoulders. Such was the dress worn in the house, made of some material both heavy and flexible, generally wool, sometimes flax—" the greatest luxury of woman," says an ancient writer. As a rule this garment was white with a coloured border, and was confined at the waist by a girdle, by which means different effects could be produced. Young girls put it round their waists, so as to make the chiton cling closely to the body and reveal its outline, while married women wore it higher, after the fashion which the Directorate borrowed from classical times. The arms were uncovered, and the feet, peeping from under the chiton, were delicately shod. "The women of Thebes," says an ancient traveller, " wear thin shoes, low and narrow and red,—they are so well laced that their feet look almost bare." The figurines of Tanagra have yellow boots with red soles.

Such was the indoor dress of Greek women, but this somewhat scanty and careless attire was neither warm nor modest nor elegant enough for the street. When they wished to dress for a walk, they put on over the chiton the himation, a large rectangular piece of stuff, sometimes called peplos and sometimes calyptra. It is difficult to define the difference between these terms, the more so that the Greeks were no less fond of new fashions than we are, and fashion is always a changeable thing. It seems, however, that the calyptra was smaller and finer, and the peplos broader and heavier.

The most important point, however, was to know how to arrange gracefully this piece of stuff, which was between four and five feet broad, and twice or two and a half times as long, sometimes white and sometimes pink, as at Tanagra, or bordered with purple or black stripes. The ways of wearing it were infinitely varied. "Every year," says M. Rayet, from whom most of our details are derived, "evidently introduced a new fashion, every change of temperature and every circumstance of life gave rise to a different arrangement, and every woman had her preference, which varied from hour to hour. If the weather were warm and she wished to be at her ease, a Greek woman would allow her calyptra to fall down behind to her waist, only keeping it up by her folded arms and letting the ends fall at either side, or taking up one of these ends, she would throw it carelessly over her left shoulder. It was then nothing more than an elegant scarf, an excuse for graceful attitudes." If she wished to drape herself more closely, she placed the piece of stuff on the top of her head, and threw the right end over the left shoulder so as to let it hang down behind; the garment clung close to the breast and left one hand at liberty. Sometimes the lower part of the face was hidden by the piece of stuff: this was the ordinary fashion among Theban women. "The part of their himation," says an ancient writer, "which forms a veil over their heads, is arranged in such a way that the face is reduced to the size of a small mask,—the eyes only are uncovered, all the rest is hidden under the mantle."

We can understand that with these broad pieces of stuff women of fashion could produce very pleasing combinations and a very harmonious disposition of folds. To arrange her drapery well was a matter of the first importance for a woman in society, and those who did not succeed in so doing were very much ridiculed. They were treated as rustics; and a poet says concerning them, "One of the marks by which we know an ill-bred person in the street is her ungraceful way of walking. What hinders you from walking well? We put no tax upon it, and it costs nothing; those who possess this advantage do themselves honour and give pleasure to the passers-by; if you are wise you should always try to gain it."

We must admit, moreover, that there was some merit in turning a commonplace piece of stuff to account, for as it was impossible to buy a complete and well-fitting costume from a fashionable dressmaker, the success of the toilette depended on the good taste of the wearer, and a well-dressed woman was necessarily an intelligent one.

Another very important feature in the toilette was the style of dressing the hair, and it is strange to see how fashion in its changes always comes round to the same point. The coiffure of a fashionable lady of Tanagra would do no discredit to a Parisian *élégante*, and it would often be easy to tell its modern name. "The three styles we meet with most frequently," says Rayet again, "are very simple; in the first, the hair is carried up to the top of the head, and there tied with a fillet so as to form a kind of tuft or bunch. In the second, the whole mass is divided by a parting on the top of the head, each half is arranged in ringlets and then carried back to the nape of the neck and there made into a kind of curl. In the last style, the hair, drawn to the back of the head, is kept up by a handkerchief, the two ends of which are fastened together on the top of the head." By the side of these simple and graceful coiffures we find, especially among singers and dancers, others which are more showy and complicated. There are enormous chignons, curls and crimps, flowers and garlands, one wears a lofty golden diadem on the front of her head, while another keeps her hair in place by a net of ribbons and gold thread, others again have gilded wreaths twined among their hair and long metal hairpins supporting the skilfully constructed pile, built up with much aid of curling-tongs and cosmetics. We can imagine the importance, among these people, of the hair-dressers who erected these marvellous structures; their shops, always full of dandies and idlers, were the rendezvous of all the newsmongers in the town. All day long they discussed the current news of the town and important political affairs, and the master of the shop led the conversation with an inexhaustible flow of words. King Archelaus is reported to have uttered a witty repartee *à propos* of this habit. A garrulous hairdresser asked him once, as he wrapped the cloth round him, "How must I cut your hair, your Majesty?"

"In silence," answered the King.

A few more accessories completed the toilette of a woman of fashion. On her head she placed a round hat, almost flat but rising to a peak in the centre, in her hand she took a fan, shaped like a lotus leaf, and generally painted a pale blue, she fastened golden ornaments on arms and hands, finally, paint and antimony embellished her face, and certain skilful mixtures gave her hair a beautiful golden-red hue something like Venetian yellow. Then she might appear in public, with quick, light step, her mirror at her side, in the delicate grace and studied daintiness of her attire.

In this hasty notice we can only bestow a passing mention on many other figurines equally pretty and graceful—such as the jugglers performing their varied feats, musicians swelling their cheeks as they blow the double flute, or playing the harp and the tambourine, dancers executing their steps to the sound of castanets and drums, and all that world of beauties, with smiling faces and saucy looks and exquisitely charming dress. The Greek genius in the fourth century took delight in these marvels of taste and fancy and piquant grace, and it is by these unrivalled beauties, and by their thoroughly modern style, that these charming figurines from the burial-grounds have bewitched our own age and won a deserved fame which the humble coroplasts of Tanagra in Boeotia can never have anticipated.

THE END.

INDEX.

Achilles, 17, 71, 78.
Acraephiae, 194, 196.
Acropolis of Athens, 24, 86, 89, 95, 97, 104.
   marbles, 89-93, 99-104, 107-110, 121, 126, 189, 287, 326.
   museum, 99, 101, 114, 125.
Actium, 199.
Aeginetan school, 118-21, 209, 210, 286.
Aegisthus, 12, 13.
Aeneas, 78, 130.
Action, 253.
Agamemnon, 7, 12, 19-21.
Agasias of Ephesus, 191.
Agrae, 313.
Alcamenes, 106, 256, 262, 267.
Alcibiades, 253, 313.
Alexander, 151, 154, 156, 195, 369.
  age of, 386, 392.
Alpheus, 211, 218, 226, 257, 259.
Alxenor of Naxos, 120.
Amazons, 277, 355.
Amphictyonic Council, 195.
Angelion, 185.
Antenor, 85, 120, 125.
Antiochus Epiphanes, 146.
Antoninus, 336, 337.
Apelles, 368.
Aphrodite, 71, 146, 149, 287, 356, 373, 378.

Apollo, 79, 130, 201.
  Delian, 130, 134.
  revenues of, 175.
  temple of, 137, 145, 170.
  Maleatas, 336, 351.
  Ptoïos, 195.
  temple of, 198.
  statues of, 102, 239, 145, 206, 207.
Apollos (so-called), 102, 186, 191, 199.
  Orchomenus, 202.
  Piombino, 207.
  Ptoïos, 203.
  Strangford, 207.
  Tenea, 204.
  Thera, 200.
Apolloniastai, 165.
Archaeological Society of Athens, 24, 136, 307, 362.
Archermus, 113, 120, 124, 189.
Argolis, 8, 23, 24, 28, 29.
Argonauts, 79.
Argos, 8, 11, 41, 196, 285.
Aristion, 116, 120.
Aristocles, 12.
Aristophanes, 313, 347.
Artemis, 130, 147, 190, 336, 396.
  Persian, 285.
  Tauric, 131.
Asclepius, 147, 332.
Assyria, 22, 31, 33, 383.
Astarte, 32, 71, 165.

Athena,—
   head of, 88, 118.
   of Endoeus, 87, 94.
   Polias, 107, 351.
   Promachos, 102, 119.
   temple of, 95, 105.
Athens, 77, 80, 133, 151, 159, 163, 196.
Atreus, 12.
   treasury of, 10, 13.
Atlas, 272.
Attic school, 83, 108, 120, 124, 355.

Boeotia, 24.
Boeotian League, 196.
   school, 201, 204, 205, 207.
Bronze, Delian, 142, 169.
Burial customs, 17, 364.
Byzantines, 12, 23.
   buildings, 42, 45, 216, 295.
Byzantium, 160, 216.

Cadmus, 27.
Calamis, 106, 232, 361
Callon, 120.
Callonides, 120.
Canachus, 205, 208.
Carapanos, 66, 73, 74, 77.
Carians, 28, 36.
Cavvadias, 87, 332.
Cecrops, 27.
Centaurs, 256, 260, 354.
Ceryces, 308.
Cesnola, 94.
Chaeronaea, 230, 360.
Chaldaea, 22, 32.
Chios, 84, 120, 123, 124, 186, 189.
Cimon, 97, 248, 249.
Cladeus, 225, 257, 259.
Clement of Alexandria, 313, 318.
Clytemnestra, 12, 19.
Cnidus, 287, 339.
Cnossus, 38, 39.
Cora, 295, 378.

Corcyra, 76.
Corinna, 361.
Corinth, 8, 146, 197.
Cos, 339.
Crete, 24, 36, 38, 120, 133, 200.
Crimea, 23, 136.
Critius, 85, 120.
Curtius, 222.
Cyclopean buildings, 10, 15, 16, 41.
Cynisca, 241.
Cynthus, 136, 137, 149.
Cyprus, 24, 38, 94.
Cypselus, 228.
Cyrene, 339.

Daedalus, 85, 107.
Danaus, 27.
Delian league, 152.
Delos,—
   excavations, 136.
   history, 150-164.
   religious importance, 130, 337.
   statues, 185-193, 199.
   trade, 164-170.
Delphi, 79.
Delphic oracle, 302.
Demeter, 295, 367, 376.
Dione, 69, 70, 71, 77.
Dionysus, 147, 288, 312, 373, 376, 379.
Diophantus, 354.
Dipoenus, 200.
Dipylon, 35, 61.
Discobolus, 250.
Discophorus, 117.
Dodona,—
   games, 79.
   oracles, 74-80.
   situation, 65.
Dorian invasion, 18, 24, 40, 61.
Dörpfeld, Dr., 45.
Dress, 91, 397.

Echo portico, 231.
Egypt, 17, 22, 29, 30, 31, 33, 69, 155, 157, 284, 383.

INDEX. 405

Electra (play), 13, 249.
Eleusis, 295.
  buildings, 323.
  statues, 107, 326.
Elis, 213, 255.
Endoeus, 85, 120.
  Athena of, 87, 94, 108, 118.
Engraved stones, 25, 38.
Epaminondas of Acraephiae, 196.
Epidaurus, 332.
  buildings, 334.
Epirus, 65, 66, 72.
Erechtheum, 87, 107.
Erisychthon, 130.
Eubuleus, 329.
Euenor, 120.
Eumolpidae, 308.
Eumolpus, 301.
Euripides, 13.
Euthycles, 120.
Ex-votos,—
  Dodona, 67, 77.
  Mount Ptoion, 95.
  Olympia, 284.

Florentine art, 82, 114.
French School of Archaeology, 136, 141, 198, 363.

Games, 79, 196, 235.
Gauls, 23.
Geranos, 133, 147.
Goths, 216.

Hadrian, 302, 305.
Halicarnassus, 271, 355.
Hallstadt, 23.
Hecate, 336, 356.
Hegesias, 85.
Helios, 297, 351.
Hellanodicae, 231, 237, 242, 252.
Hera,—
  festival of, 254.
  head of, 286.
  of Samos, 84, 121.
  temple of, 229, 281.

Heracleistai, 148, 165.
Hercules, 41, 78, 212, 243.
  and Atlas, 272.
  and Cretan Bull, 221.
  and Hydra, 89, 99.
  and Triton, 99.
  and Typhon, 100.
  labours of, 256, 272.
Hermaistae, 148, 165.
Hermes of Praxiteles, 288.
Herodes Atticus, 231.
Herodotus, 29, 69, 70.
Heruli, 23.
Hieropoioi, 178, 311.
Hippocrates, 347.
Hippodamia, 212, 259.
Hissarlik, 6, 38, 39, 40.
Holleaux, M., 198.
Homeric hymns, 132, 151, 300.
Homeric poems, 6, 11, 12, 15, 17, 22, 25, 29, 44, 52, 71.
Homolle, M., 137, 141.
Horns, altar of, 130, 147.
Hygieia, 336, 351, 356.
Hymns, Homeric, 132, 151.

Iacchus, 315, 329.
Ialysus, 38, 39.
Ictinus, 303, 325.
Inscriptions, 73, 150, 198, 322, 341.
Iphitus, 212.
Iron, 25.

Janina, 65.

Karditza, 194.
Kharvati, 8.
King archon, 309, 310.

Lapiths, 256, 260, 354.
Larissa, 63.
Lefsina, 295.
Leonidaeum, 234.
Leto, 130, 146.
Libon, 256.
Libya, 70.

Lion gate, 10, 13, 14.
Lucian, 80.
Lycurgus, 303, 311.
Lysippus, 290, 292, 368.

Macedonia, 155, 158.
Mantinea, 196.
Masks, 18, 34.
Megara, treasury of, 287.
Megaron, 54.
Menidi, 24, 36, 56, 59, 60.
Messenians, 268.
Methe, 334, 373.
Metroum, 230.
Micciades, 189.
Milo of Croton, 245, 247.
Minos, 36.
Mithridates, 140, 168.
Morea, expedition to, 23, 220.
Moschophorus, 87, 95, 115, 122, 123.
Mycenae, 34, 38, 39, 46, 52, 56, 59.
  civilisation, 26, 35, 37, 40, 59.
  lower city, 25.
  palace, 25, 46, 52, 56, 59.
  situation, 8.
  tombs, 15, 19.
Mycono, 129.
Myrina, 363, 374.
Myron, 106, 244, 250, 293.
Myrtilus, 257.
Mysteries, greater, 316.
  hall of, 303, 324.
  lesser, 313.
  of Samothrace, 314.

Nausicaa, 395.
Naxos, 139, 187, 199.
Nereids, 271, 355.
Nero, 198.
Nesiotes, 85.
Nicandra, 187.
Nicias, 134.
Nysa, 295, 378.

Oenomaus, 213, 257.
Onatas, 120, 232.
Oracles,—
  Delphic, 79, 238.
  Dodonaean, 72.
  Lebadaea, 196.
  Ptoian, 195.
Orchomenus, 7, 38, 56, 59.
  Apollo of, 199, 201, 202.
Orestes, 12, 20, 78.
  (play), 13, 15.
Origen, 313, 318.
Oriental influence, 8, 27, 31, 37, 39, 59-61, 285.

Paeonius, 233, 256, 262, 267, 269, 355.
Palaces, Homeric, 25, 44, 46, 52, 58, 62.
Panaenus, 276.
Pancratium, 236, 247.
Panhellenes, 302, 306.
Parthenon, 86, 88.
Paros, 120.
Pausanias, 13, 21, 139, 215, 232, 248, 260, 276, 281, 306.
Pausias, 334.
Pediment sculptures,—
  Athens, 88, 89, 99, 102, 123.
  Delos, 192.
  Megara, treasury of, 287.
  Olympia, 257, 260.
Pelasgic walls, 104.
Pelopium, 229.
Peloponnesian school, 207, 273.
Peloponnesus, 196, 220.
Pelops, 12, 27, 28, 211, 212, 228, 257, 258.
Penrose, 45.
Pentathlum, 236, 249.
Perdicovrysi, 198.
Pericles, 86, 95, 104, 303.
Persians, 86, 95, 97, 102, 105, 134, 153, 195, 325.
Pharaohs, 30, 31.
Pheidolas, 249.
Pherenice, 241.

Phidias, 106, 119, 234, 275, 280.
Philip, 154.
Philip V., 148.
Philippeum, 230.
Philo, 120.
　of Eleusis, 305.
Philostratus, 81.
Phoenicia, 22, 28, 29, 31, 33, 36, 59, 60, 61, 70, 284.
Phrygia, 12, 28, 36.
Pindar, 194, 235, 361.
Pirithous, 260.
Pisistratus, 95, 98, 105, 118, 124, 303.
Plato, 22.
Plutarch, 319, 320.
Pluto, 295, 300.
Polycleitus, 334, 336.
Porinos Oikos, 144, 181.
Poseidoniastai, 148, 165.
Pottery (Mycenaean), 25, 34, 61.
Praxiteles, 288, 290, 330, 368, 381.
Propylaea, 52, 306, 323, 336.
Ptolemies, 157, 180.
Pygmalion and Galatea, 107.
Pythia, 79.

Rameses II., 21, 30.
Rhodes, 24, 38, 155, 156, 159, 339, 383.
Roman period, 196.
Romans, 161-3, 166, 238.

Samos, 84.
Santorin, 24, 38, 39.
Schliemann, Dr.,—
　at Mycenae, 14.
　at Tiryns, 44.
　excavations, 6, 7.
　life, 2.
Schliemann, Mrs., 4, 11.
Scopas, 271, 355, 368.
Selli, 69, 71.
Serapis, 136, 149, 165, 185.
Sicyon, 84, 218.
Skimatari, 360, 362.
Skyllis, 199.

Slavonic invasion, 12, 218.
Sophocles, 13, 322.
Sparta, 79, 84, 153, 238, 345.
Spata, 24, 37, 39, 60, 61.
Sphacteria, 268.
Sterope, 257, 259.
Syra, 128, 169.
Syrens, 388.
Syria, 71, 72, 155.
Sword-blades, 32.

Tanagra, 17, 360.
Tectaeus, 185.
Telesphorus, 357.
Tenea, Apollo of, 199, 204.
Terra-cottas, 283, 366.
Thebes, 195, 359, 360.
Themis, 336.
Theodorus of Samos, 120, 125.
Theodosius, 216.
Theoriai,—
　Delos, 133, 159.
　Olympia, 252.
Thera, Apollo of, 199.
Theseum, 260, 274.
Theseus, 130, 133, 147, 260, 277.
Thespiae, 359, 361.
Thessaly, 24, 38, 63.
Thrasymedes, 334, 355.
Thucydides, 12.
Tinos, 140, 146, 169.
Tiryns, 25, 38, 39, 41.
　citadel, 46.
　galleries, 48.
　paintings, 56.
　palace, 45, 52, 59.
　walls, 42, 47.
Tomaros, 66, 72.
Tombstones, 15, 16, 18.
Treasuries, 230.
Tricca, 336.
Triptolemus, 299, 322, 328.
Troezen, 343.
Troy, 6.
Turks, 11, 12, 218.
Typhon, head of, 100.
Tyrannicides, 85, 125.

Ulysses, 29, 52, 78, 130.

Vaphis, 38.
Veli Pacha, 11.
Victories, winged, 113, 124, 188, 355.
Victory of Paeonius, 233, 268, 270, 355.
Victory of Samothrace, 271.

Winckelmann, 219.

Zanes, 232.
Zeus,—
  altar of, 228.
  archaic, 286.
  Dodonaean, 66, 71-80.
  Herceius, 53.
  of Phidias, 217, 276.
  Otricoli, 279.
  Polieus, 147.
  temple of, 227, 255.
  Verospi, 278.

H. GREVEL & CO.'S PUBLICATIONS.

FINE ARTS.

THE CLASSICAL PICTURE GALLERY

FOR 1893: Monthly Magazine of Reproductions from the Art Galleries of Europe. Each Part 1s., containing Twelve Plates, 4to. Annual Subscriptions, 14s. post free.

"Judged by their price, the plates are little short of wonderful. They will be of no little value to the serious student of art, who will be able by their means to compare types and expressions, and refresh his memory in various ways."—*Saturday Review.*

"It is worth the attention of those who wish to make a collection of illustrations of the works of great painters at a small cost." *Builder.*

"All the principal galleries in England and on the Continent have been laid under contribution, and the masterpieces of the great painters are admirably reproduced in process at so cheap a rate that each number of the Classical Picture Gallery is given for one shilling. This collection is by far the best of the cheap editions of copies from the Old Masters."—*Graphic.*

"The reproductions are marvellously cheap, and they certainly give a far better idea of the pictures represented than any publication we have seen. To those who have an interest in art, to whom pictures in foreign galleries are mere names, the work will be most valuable; nor will it meet a less cordial welcome from those who desire to recall what they have seen."—*Birmingham Daily Post.*

"The idea is well carried out, and it should prove popular."—*The Art Journal.*

THE CLASSICAL PICTURE GALLERY.

Annual Volume for 1890 and 1892, containing each 144 Plates. With Biographical Notices of the Artists, and a Complete Index. 4to, cloth extra, gilt top. Each Volume £1 1s. (Vol. for 1891 is out of print.)

H. GREVEL & CO., 33, KING STREET, COVENT GARDEN, LONDON, W.C.

Excursions in Greece.

FINE ARTS—*Continued.*

PORTFOLIO FOR COLLECTING THE MONTHLY PARTS OF THE CLASSICAL PICTURE GALLERY. With flaps, cloth gilt. 3s.

DIEGO VELAZQUEZ AND HIS TIMES. By CARL JUSTI, Professor at the University of Bonn. Translated by Professor A. H. KEANE, and Revised by the Author. With 52 magnificent Woodcuts, an Etching of Velazquez' Own Portrait by Forberg, and a Plan of the Old Palace at Madrid. 1 Vol. Royal 8vo, Roxburgh. £1 1s.

"No better example could be given of the growth of historical knowledge during the past thirty years than is to be found in Carl Justi's recent book on 'Velazquez,' well translated by Professor A. H. Keane. The amount of research and of real observation which his book shows is typical of the modern scholar."—*Times.*

"It may appear a paradox to state that no complete or adequate account of the life and works of Diego de Siloa Velazquez has been written until the masterly Biography by Professor Carl Justi, which will long be looked upon as the definitive biography of the Great Master."—*Edinburgh Review.*

"Considering the estimation in which the works of Velazquez are now held in this country, it was fitting that the monograph which Professor Justi has devoted to his life and times should be presented in an English form to the public. The book has been written with true German thoroughness, and the labour involved in amassing and digesting the enormous amount of information contained in it is beyond conception. Every picture that Velazquez ever painted seems to be enumerated. Every event in his life is carefully recorded."—*Morning Post.*

MUSIC.

RICHARD WAGNER'S LETTERS TO HIS DRESDEN FRIENDS (Letters to Theodor Uhlig, 1849-1853; Letters to Wilhelm Fischer, 1841-1859; Letters to Ferdinand Heine, 1841-1868). Translated by J. S. SHEDLOCK. With an ETCHING by C. W. SHERBORN of WAGNER'S PORTRAIT taken in 1853, and a COMPLETE INDEX. 1 Vol. Crown 8vo, cloth extra, gilt top. 12s. 6d.

"Admirers of Wagner's genius will find here a fund of information about the *vie intime* of the musician."—*Graphic.*

"All who are interested in the personal character and career of Richard Wagner —such persons may hold diverse opinions as to his art—will welcome this book as a companion volume to the English version of the correspondence between Wagner and Liszt."—*The Musical Times.*

"It is quite impossible within limited space to give a satisfactory description of the contents of this extremely interesting volume."—*Athenæum.*

"These letters are supremely interesting, for they show Wagner as he really was no one-sided representation, but the whole man."—*Daily Telegraph.*

H. GREVEL & CO., 33, KING STREET, COVENT GARDEN, LONDON, W.C.

MUSIC—*Continued.*

RICHARD WAGNER'S CORRESPONDENCE WITH LISZT FROM 1841 TO 1861.
Translated into English, and with a Preface by Dr. Francis Hueffer. 2 Vols. Crown 8vo, cloth, gilt top. £1 4s.

"Nothing more instructive with regard to the real character and relations of Liszt and Wagner has been published. Seldom has the force and fervour of Wagner's German been rendered with such accuracy and character in a strange tongue."—*Manchester Guardian.*

MANUAL OF MUSICAL HISTORY.
With 150 Illustrations of Portraits, Musical Instruments, Facsimiles of Rare and Curious Musical Works. By James E. Matthew. 1 Vol. Crown 8vo, cloth. 10s. 6d.

"This book is well written, interspersed now and again with an anecdote or amusing passage in the lives of the great composers."—*Morning Post.*

"The sketch of the later music in Germany, Italy, France, and England, brings the book up to date, and concludes a well-written, profusely illustrated, and admirably compiled volume."—*Musical News.*

"The plan of the book is admirably arranged and printed. The work can be well recommended to all who desire a musical history, giving full information in a concise and accessible form, and which is detailed without being diffuse."—*Musical Standard.*

"The author covers the entire ground of his subject from the time of St. Gregory to the present day, and serves admirably for those who wish to obtain some general idea as to the progress of the art. The facts are succinctly set forth, the statements are correct, and the many illustrations give the work a distinct value."—*Daily Telegraph.*

"The book is handsomely got up in every way."—*Queen.*

VERDI: an Anecdotic History of his Life and Works.
By Arthur Pougin. With Portrait and Facsimile. Translated from the French by James E. Matthew. Crown 8vo, cloth extra. 6s.

"It is long since a more readable and fascinating book has come under our notice. Mr. Matthew has acquitted himself excellently of the task of translation, and the well-printed volume will be heartily welcomed by musicians."—*Athenæum.*

"Mr. Matthew's well-written translation of Pougin's work would at any time be pleasant, and just now must have special charms."—*The Quarterly Musical Review.*

"The publication is very timely. It comes just when most wanted, and enables musical readers to study every stage of the fortunate career which has now reached its climax and consummation."—*Daily Telegraph.*

"This extremely interesting volume will be much appreciated in the musical world."—*Pall Mall Gazette.*

"The work has a good portrait and a good index, and for the latter the translator deserves grateful thanks."—*Publishers' Circular.*

H. GREVEL & CO., 33, KING STREET, COVENT GARDEN, LONDON, W.C.

THE STUDENT'S ATLAS OF ARTISTIC
ANATOMY, for the Use of Sculptors, Painters, and Amateurs. With THIRTY-FOUR PLATES. By CHARLES ROTH, Professor of Sculpture at the Munich Academy. Edited by C. E. FITZGERALD, M.D. 1 Vol. Fol., in cloth case. £1 5s.

"Professor Roth's well-known and highly valued plates of the human muscles and bones, with his nomenclature and descriptive notes, are worthily reproduced in this complete portfolio."—*Athenæum*.

"The plan of the Atlas, which has been considerably enlarged from its original form, is very simple and convenient. The two first plates represent the bony framework of the body, and the eight following ones the muscles in action. The figure plates are carefully drawn and admirably printed, and each is followed by an explanatory plate, in which the parts depicted are numerated, while notes are added directing attention to those points which are of especial interest to artists."—*Daily Graphic*.

"It is a bold experiment to attempt to combine at once the classical, natural, and anatomical elements in drawing from the nude, but in this the author has succeeded remarkably well. The limitation to a single pose prevents all confusion, and very much facilitates anatomical knowledge, whilst the addition of separate sketches at the end of the atlas furnishes an opportunity for more detailed study of the limbs and head. The plates are admirably executed, the engraving being smooth and even, and at the same time sufficiently vigorous in the contrast of light and shade. Dr. Fitzgerald's translation of Professor Roth's work is altogether a valuable aid to the study of artistic anatomy, and in itself a work of art."—*Lancet*.

THE HUMAN FIGURE: its Beauties and
Defects. By ERNEST BRÜCKE, Emeritus Professor of Physiology in the University of Vienna, and formerly Teacher of Anatomy in the Academy of Fine Arts at Berlin. With 29 Illustrations by HERMANN PAAR. Edited, with a Preface, by W. ANDERSON, Professor of Anatomy to the Royal Academy of Arts, London. Crown 8vo, cloth. 7s. 6d.

"An exceedingly valuable work for sculptors, painters, and all such as, possessing some knowledge of anatomy, are engaged in artistic pursuits, and one which may likewise be read with pleasure and profit by the comparatively uninitiated. Written in a clear, concise style, equally removed from pedantry as from superficiality, Herr Brücke combines an artist's sentiment with the anatomist's keen eye, and his work is evidently the result of thoughtful study. The artist, he tells us, should be able to detect the faults of the human frame as easily as the horse connoisseur recognises the defects of an animal."—*Times*.

"The work is primarily addressed to artists, but all those who are interested in the artistic representation of the human body will find it a very interesting and suggestive work."—*Edinburgh Medical Journal*.

"No book could be more welcome to the English artist and anatomist."—*Lancet*.

"We can strongly recommend it both to the art-student and to the full-fledged artist."—*Saturday Review*.

"We cannot single out the numerous points of interest touched upon."—*Dublin Journal of Medical Science*.

"It is long since we have been called upon to review a more thorough, learned or interesting book upon an attractive and important subject."—*Athenæum*.

H. GREVEL & CO., 33, KING STREET, COVENT GARDEN, LONDON, W.C.

ARCHÆOLOGY.

MANUAL OF ORIENTAL ANTIQUITIES:
including the Architecture, Sculpture, and Industrial Arts of Chaldæa, Assyria, Persia, Syria, Judæa, Phœnicia, and Carthage. By ERNEST BABELON, Librarian of the Department of Medals and Antiques in the Bibliothèque Nationale, Paris. Translated and enlarged by B. T. A. EVETTS, M.A., of the Department of Egyptian and Assyrian Antiquities, British Museum. With 241 Illustrations. Crown 8vo, cloth extra, gilt top. 10s. 6d.

"The 'Manual of Oriental Antiquities,' which takes deservedly a high position both for the general accuracy of its statements and excellent character of its illustrations."—*Athenæum.*

"This carefully written and beautifully illustrated book forms a worthy companion to the Egyptian Archæology of M. Maspero included in the same series."—*Manchester Guardian.*

"M. Babelon's work has already won such a high measure of praise from the students of Oriental archæology that its place is assured among modern authorities on the art and culture of the ancient civilised nations of Western Asia."—*Academy.*

"This book will be most valuable and useful, not only to students of Oriental archæology, but to every one who is interested in the early beginnings of architecture, sculpture, and the industrial arts. The quantity of information contained in one small volume is wonderful, embracing the art history of Chaldæa, Assyria, Persia, Syria, Judæa, Phœnicia, and Carthage. By the study of the illustrations alone a clear idea can be gained of the different characteristics of these nations."—*Spectator.*

MANUAL OF GREEK MYTHOLOGY IN RELATION TO GREEK ART.
By MAXIME COLLIGNON, late Member of the Ecole Française, Athens. Translated and Enlarged by JANE E. HARRISON, Author of "Myths of the Odyssey," "Introductory Studies in Greek Art," etc. With 140 Illustrations. Crown 8vo, cloth extra, gilt top. 10s. 6d.

"This is a good book. It fills a gap in our literature, doing for Greek mythology much the same service which Mrs. Jamieson rendered to Christian hagiology."—*Academy.*

"The gradual development of the type of each god according to the general principles which govern the formation of types in art is historically considered. Miss Harrison's translation does not read like a translation, but like an English book, and we know no higher praise than this."—*Saturday Review.*

H. GREVEL & CO., 33, KING STREET, COVENT GARDEN, LONDON, W.C.

ARCHÆOLOGY—*Continued.*

MANUAL OF ANCIENT SCULPTURE.

By PIERRE PARIS, formerly Member of the Ecole Française, at Athens. Edited and Augmented by JANE E. HARRISON, Author of "Myths of the Odyssey," "Introductory Studies in Greek Art," etc. With 187 Illustrations. Crown 8vo, cloth extra, gilt top. 10s. 6d.

"It is written in a felicitous and interesting style, and its illustrations reproduce in an accurate manner the best of the sculptured treasures which are referred to in the text."—*Scotsman.*

"A book of standard merit, it has received additions of great value from the accomplished archæologist through whose hands it has passed in the process of adaptation to the needs of English readers."—*Spectator.*

"Mr. Paris' admirable manual, which owes much more than its English dress to its accomplished editor, gives us within the narrow compass a lucid and compendious survey of the sculpture of the ancient world."—*St. James' Gazette.*

MANUAL OF EGYPTIAN ARCHÆOLOGY.

By Professor G. MASPERO, D.C.L., Oxon. English Edition, with Notes, by AMELIA B. EDWARDS, Ph.D., LL.D. With 299 Illustrations. New Edition, revised, with a Complete Index. Crown 8vo, cloth extra, gilt top. 10s. 6d.

CONTENTS.—Chapter I. Civil and Military Architecture. Chapter II. Religious Architecture. Chapter III. Tombs. Chapter IV. Painting and Sculpture. Chapter V. The Industrial Arts.

"It is a marvel of erudition and condensation. It sums up the long results of thousands of years of Egyptian civilisation in language precise enough to make the work a handbook for the specialist, and popular enough to insure its becoming a guide to the antiquarian lore of the country for travellers in Egypt."—*Scotsman.*

MANUAL OF ARCHÆOLOGY:

containing an Introduction to Egyptian and Oriental Art, Greek, Etruscan, and Roman Art. With 114 Illustrations. By TALFOURD ELY, Member of the Councils of the Society for the Promotion of Hellenic Studies. 1 Vol. Crown 8vo, cloth. 6s.

"A most admirable gift to those who take an intelligent interest in ancient art would be the 'Manual of Archæology,' by Talfourd Ely."—*Graphic.*

"We know of no such complete and concise handbook to ancient art, for Mr. Ely, beginning with Egypt, the mother of the sciences, takes us through the art of Chaldea and Assyria to that of Greece and Rome."—*St. James' Gazette.*

H. GREVEL & CO., 33, KING STREET, COVENT GARDEN, LONDON, W.C.

ARCHÆOLOGY—*Continued.*

PICTORIAL ATLAS TO HOMER'S ILIAD
AND ODYSSEY. 36 Plates, containing 211 Illustrations from Works of Ancient Art. With a complete Descriptive Text, and an Epitome of the Contents of each Book, for the use of Schools and Students of Literature and Art, by Dr. R. ENGELMANN and Professor W. C. ANDERSON, Firth College, Sheffield. 1 Vol., 4to, cloth. 10s. 6d.

"Every school library should possess this book. It is of the very highest interest."—*Spectator.*

"An excellent educational idea. The editors regard the Homeric poems as a secular Greek Bible, the illustrations to which not only make the Iliad and the Odyssey more intelligible to the Greek student, but supply a clue to the sacred mysteries of ancient art and literature."—*Daily Telegraph.*

"The atlas is thus not only useful as an aid to the study of the poems, but available also for the purposes of a student of ancient art as a thing by itself."—*Scotsman.*

"It ought certainly to prove of service to teachers who wish to render 'Homer' intelligible to their younger pupils."—*Standard.*

"The compilers have gone through 'Homer' book by book, and whenever they found an opportunity they have reproduced a bust, or mural painting, or a group of statuary illustrating and elucidating the text."—*Graphic.*

"Mr. Anderson has done valuable work in preparing this English edition of Engelmann's well-known 'Homeric Atlas.' The additions are all clear, and have been admirably planned."—*Classical Review.*

"It is a book which every one who loves his 'Homer' should study and keep at his elbow."—*University Correspondent.*

OLYMPOS: Tales of the Gods of Greece and
Rome. By TALFOURD ELY. With 47 Woodcuts and 6 full-page Photographic Plates. 1 Vol. Crown 8vo, cloth. 10s. 6d.

"In 'Olympos' Professor Talfourd Ely furnishes a book which should be found useful by a large number of readers of the present day. The author has set himself to provide an account, which shall be at once systematic and readable, of the chief Greek and Roman deities; accurate according to the light of modern researches and theories, and popular enough in style for general reading—a kind of Lemprière up to date. One feature of the book should be especially useful to young readers. Zeus and Jupiter, Hera and Juno, Poseidon and Neptune, and the rest, have parallel but distinct accounts given to them, so that the difference of origin of Greek and Roman deities is made clear."—*St. James' Gazette.*

"This is a really excellent book of its kind."—*The Spectator.*

"The book is most pleasantly written, and is the fruit of laborious research among the best available authorities on the subject."—*Echo.*

"Makes a capital gift book."—*Scotsman.*

H. GREVEL & CO., 33, KING STREET, COVENT GARDEN, LONDON, W.C.

ARCHÆOLOGY—*Continued.*

EXCURSIONS IN GREECE to recently Explored Sites of Classical Interest: Mycenæ, Tiryns, Dodona, Delos, Athens, Olympia, Eleusis, Epidauros, Tanagra. A Popular Account of the Results of Recent Excavations. By CHARLES DIEHL, late member of the French Schools of Rome and Athens, and Professor at the University of Nancy. Translated by EMMA R. PERKINS, Headmistress of the Thetford Grammar School for Girls. With an Introduction by REGINALD STUART POOLE, LL.D., Professor of Archæology at University College, London; Corresponding Member of the Institute of France. Illustrated. Crown 8vo, cloth, price 7s. 6d.

The aim M. Diehl has set before himself in this book is—discarding the technicalities of Archæology and leaving on one side its more obscure problems—to present in a clear and simple form the chief results of the excavations which, within the last twenty years, have done so much to enlarge our knowledge of the art, the life, and the religion of the Greeks. The work is thus a summary of the latest results of Archæology in Greece, and the only complete and succinct view of the whole field, with needful plans of each site.

THE MASTER OF TANAGRA: an Artist's Story of Old Hellas. By ERNST VON WILDENBRUCH. Translated from the Seventh German Edition. Illustrated with 25 Tanagra Figures. Crown 8vo, cloth extra, gilt top, 5s.

This is a novel.

"In 'The Master of Tanagra' the life of old Greece is reconstructed with a realism that gives a correct reflex of the times in a language that is both elegant and picturesque."—*Morning Post.*

BIBLIOGRAPHY.

THE BOOK: its Printers, Illustrators, and Binders, from Gutenberg to the Present Time. By HENRI BOUCHOT, of the National Library, Paris. With a Treatise on the Art of Collecting and Describing Early Printed Books, and a Latin-English and English-Latin Topographical Index of the Earliest Printing Presses. Containing 172 Facsimiles of Early Typography, Book Illustrations, Printers' Marks, Bindings, numerous Borders, Initials, Head and Tail Pieces, and a Frontispiece. 1 Vol. Royal 8vo, vellum cloth. £1 1s.

"Beginning with the Block Books, which anticipated by a few decades the discovery of Printing, this work gives an account of the rise and progress of Printing, the dispersion over Europe of the German printers, the growth of Book Illustration, of the Binder's Art, and all similar matter down to the present day."—*Athenæum.*

BIBLIOGRAPHY—*Continued.*

MANUAL OF BIBLIOGRAPHY: being an Introduction to the Knowledge of the Book, Library Management, and the Art of Cataloguing. With a Latin-English and English-Latin Topographical Index of the Early Printing Centres. With 37 Illustrations. By WALTER T. ROGERS, Inner Temple Library. New Edition, Enlarged. 1 vol. Crown 8vo, cloth. 5s.

"Contains much rare and curious information."—*Morning Post.*

"To this little work the printer, the bookbinder, the artist, the author, the librarian, the bibliophile, will all turn with interest, and, we venture to predict, will not be disappointed. It describes 'The Invention and Progress of Printing,' treats elaborately of 'The Book,' 'The Ornamentation of the Book,' 'The Library and the Catalogue,' and kindred topics. A useful list of books of reference, a glossary, and an index are added. The book is nicely got up, and the illustrations add to its beauty and value."—*Publishers' Circular.*

"It gives in a pleasant way a great deal of information about the invention and progress of printing, about rare books, matters of abbreviation, illuminated capitals, wood engraving, steel engraving, heliotype, etc., and a great many other subjects which touch the hearts of book lovers and book maniacs. It also contains many illustrations of mediæval and modern printing, binding, and decoration."—*Echo.*

LACE BOOK.

MANUAL OF EMBROIDERY AND LACE: their Manufacture and History from the remotest Antiquity to the present Day. By ERNEST LEFÉBURE, Lace Manufacturer and Administrator of the Musée des Arts Decoratifs, Paris. Translated and Enlarged, with Notes and New Designs, by ALAN S. COLE, of the South Kensington Museum. With 156 Illustrations. Crown 8vo, cloth extra, gilt top. 10s. 6d.

"A book which is not only a comprehensive, well-arranged, and trustworthy history of a delightful art, but a book which is pleasant to see and pleasant to read, well written and well edited. Its manner of production does much credit to the publishers."—*Academy.*

FENCING AND GAMES.

SKAT: AN ILLUSTRATED GRAMMAR OF the German Game of Cards called Skat, with a Bibliography on the Subject. By ERNST EDUARD LEMCKE. 1 Vol. 8vo, cloth. 4s. 6d.

"The game offers so great a variety of combinations that, though there is almost unlimited scope for skill or good play, still there is a sufficient element of chance to maintain the interest to the other hands, however apparently bad the cards they hold, as opposed to the player. The student who has patience to go carefully through Mr. Lemcke's book will be able to master the principles and rules of the game."—*Morning Post.*

H. GREVEL & CO., 33, KING STREET, COVENT GARDEN, LONDON, W.C.

FENCING AND GAMES—*Continued.*

THE SWORDSMAN: a Manual of Fence for the Foil, Sabre, and Bayonet. With an Appendix consisting of a Code of Rules for Assaults, Competitions, etc. By ALFRED HUTTON, late Capt. King's Dragoon Guards, Author of "Cold Steel," "Fixed Bayonets," etc. With 42 Illustrations. 1 Vol. Crown 8vo, cloth. 3s. 6d.

50 copies printed on Whatman paper, bound in vellum, at 10s. 6d.

"Captain Alfred Hutton, a well-known authority on the subject, has compiled a book, which should prove of much service to the beginner, and which also contains many useful hints to the more accomplished performer."—*Morning Post*.

"Books of the kind are rare in English, and as Captain Hutton is a skilled master, and his method carries authority, the book deserves a welcome from all who are interested in the maintenance of good swordsmanship."—*Scotsman*.

"We do not know a work that is so full of clear explanation and sound instruction on the art of fencing."—*Court Journal*.

"Captain Hutton, a well-known authority on all connected with the use of cold steel, has here dealt with another branch of the subject, which he has further explained by many admirably drawn illustrations."—*Journal of the United Service Institution*.

"This little treatise should be read by all who take any interest or who desire to gain a practical knowledge in the art of fence, an art which has, unfortunately, been much neglected of late years. Captain Hutton deserves the gratitude of all admirers of the *arme blanche*, not only for his able advocacy of the art, but also for his excellent exposition of its practice. The book, which is replete with valuable lessons, contains forty-two illustrations, and will be found a most valuable companion to his other works, 'Cold Steel,' 'Fixed Bayonets,' etc."—*United Service Gazette*.

OLD SWORD PLAY. By Capt. ALFRED HUTTON, Author of the "Swordsman," "Cold Steel," and "Fixed Bayonets," etc., with 58 Illustrations after Alfieri, Angelo, Di Grassi, Liancourt, Marozzo, De la Touche, Weischner, and other celebrated Fencing Masters, containing a Series of Studies of the Swordsmanship of the 16th, 17th, and 18th Centuries, embracing the Twohand Sword, Rapier, and Dagger, Broadsword and Buckler, "Case of Rapiers," Early Small Sword Play, etc. Royal 8vo. Superfine Dutch paper. Vellum cloth, extra. Limited to 300 copies. Buckram, £1 1s.

"Captain Hutton has compiled and arranged the lessons in his book so as to make the antique methods accessible to the student without the labour of searching through many ancient volumes."—*Graphic*.

"Captain Hutton's clear and concise treatment of this curious form of Sword Play, illustrated as it is by plates from Marozzo and Di Grassi, is extremely interesting. This latest contribution to the literature of fencing should not be neglected by any one interested in that fine art."—*Illustrated London News*.

"The plates are superbly reproduced, and form a most valuable collection."—*Saturday Review*.

H. GREVEL & CO., 33, KING STREET, COVENT GARDEN, LONDON, W.C.

MEMOIRS.

A HANOVERIAN-ENGLISH OFFICER A HUNDRED YEARS AGO. Memoirs of Baron Ompteda, Colonel in the King's German Legion, 26th November, 1765, to 18th June, 1815. Translated by JOHN HILL, M.A. With Portrait. Demy 8vo, cloth, price 16s.

The life of Baron Ompteda, who was born in 1765, and fell on the field of Waterloo in 1815, is full of variety, and rich in deeds and sufferings. He took part in the Coalition War in the Netherlands in 1793-4, and was engaged thenceforward to his end in the struggle maintained by England and Hanover against the French domination. On the breaking up of the old Hanoverian army, in 1803, he was one of those officers who forthwith constructed from its fragments the King's German Legion. With this corps war took him to Ireland, to Gibraltar, to Copenhagen, into the hands of the Dutch and French as a prisoner of war, to Spain (1812-14), until at last he met a hero's death at Waterloo. The work throws also many sidelights on English politics and society under George III., and teems with interesting anecdotes. Baron Ompteda was a man distinguished for striking intellectual gifts and general culture. He was intimately acquainted with Scharnhorst and Gneisenau, as the correspondence testifies.

"A very interesting life of a very interesting soldier"—GENERAL VISCOUNT WOLSELEY.

MEDICINE.

MY WATER CURE TESTED FOR MORE THAN 35 YEARS for the CURE of DISEASES and the PRESERVATION of HEALTH. By SEBASTIAN KNEIPP, Parish Priest of Wörishoven, Bavaria, with 100 Illustrations and a Portrait of the Author. Complete Copyright Edition. Enlarged by 136 pages; translated from the last (the 36th) German Edition. Entirely revised by the Author, with 50 additional Illustrations of Medicinal Plants. Post 8vo, cloth, 396 pp., price 6s.

This Edition is an accurate rendering of the Original Work, as revised by the Author himself, and is not to be confounded with another Translation, of 272 pages, price 5s., which is made from an old German Edition, and contains neither the Original Woodcuts nor the New Illustrations of Medicinal Plants.

"All Germany bears witness to the worth of his regimen."—*The Guardian*.

"Here is an author whose book has already gone through thirty-six editions in German alone. There is no form of water cure but the Bavarian pastor has something to say about. This book deals, first, with the numerous forms of water application, shawls, bandages, wet shirts, etc., next with medicines, and, thirdly, with diseases. We hope it may find a wide circulation in England."—*Echo*.

"The author expresses himself with directness and simplicity which are truly refreshing. Common sense is the prevailing element running through the book, and by common sense and observation many diseases may be averted. This volume should take its place as an honoured member of the domestic pharmacopœia of the day." *Sala's Journal.*

"The translation of this valuable work into English will have the effect of giving us a handbook to health the like of which we cannot boast. The idea of the culture of health is in a sense original, since the methods instituted by the author have in their entirety never been adopted in England. Hydropathy, it is true, has long been popular here, but not altogether on the lines of Sebastian Kneipp. His system he has given in the work under notice, a work that is at once original in conception and able in execution."—*Christian Union*.

H. GREVEL & CO., 33, KING STREET, COVENT GARDEN, LONDON W.C.

MEDICINE—*Continued.*

PLANT-ATLAS, illustrating SEBASTIAN KNEIPP'S "My Water Cure"; describing and picturing true to nature all Medicinal Plants mentioned in his books, with addition of several others frequently resorted to by the people. With 41 coloured Plates containing 69 specimens. 8vo, cloth, 10s. 6d.

Everybody is enabled, by simply consulting the "Plant-Atlas," to find out for himself whatever herb he will have to look for in woods or fields, and thus to make up, in a most pleasant way, that "Family-Medicine-Chest," recommended by Kneipp.

The description is supplied by a distinguished Botanist, and dwells on any point of importance, for instance upon the General and Special Characteristics of every Plant, its flowering time, use, occurrence, diagnostics, mode of acting, healing power, etc.

THE KNEIPP BROCHURE: containing all about the Life and Works of the celebrated Woriskofen Priest, and his Healing Method. Illustrated. Free on Application.

CORPULENCE AND ITS TREATMENT on Physiological Principles. By Dr. W. EBSTEIN, Professor of Medicine and Director of the Clinical Hospital at Göttingen. 1 Vol. 8vo, boards. 2s. 6d.

"This work is written in a thoroughly scientific spirit, and has no sign of charlatanry about it."—*Medical Times.*

GERMAN LANGUAGE.

PRACTICAL GRAMMAR OF THE GERMAN LANGUAGE. With Reading Lessons, and a German-English and English-German Vocabulary. By WILLIAM EYSENBACH. New Edition. 1 Vol. 8vo, cloth. 3s. 6d.

"It is decidedly '*practical.*' We like the gradual mode of presenting difficulty after difficulty, and the conversational tone of the exercises."—*Schoolmaster.*

DICTIONARY OF THE ENGLISH AND GERMAN LANGUAGES. By Dr. FR. KOEHLER. 30th Edition, brought up to date, by Professor Dr. H. LAMBECK. The best German Dictionary extant, even better for practical use than any of the larger Dictionaries. 1 Vol. 8vo, cloth, 7s. 6d.

H. GREVEL & CO., 33, KING STREET, COVENT GARDEN, LONDON, W.C.

www.ingramcontent.com/pod-product-compliance
Lightning Source LLC
Chambersburg PA
CBHW032136010526
44111CB00035B/588